T5-BBB-865

Power Tools and How to Use Them

New Illustrated Library of Home Improvement Volume 5

Power Tools and How to Use Them

Prentice-Hall/Reston Editorial Staff

Prentice-Hall of Canada, Ltd. / Reston Publishing Company
Scarborough, Ontario

Series contributors/ H. Fred Dale, Richard Demske, George R. Drake, Byron W. Maguire, L. Donald Meyers, Gershon Wheeler

Design/ Peter Maher & Associates
Color photographs/ Peter Paterson/Photo Design

Printed and bound in Canada.

The publishers wish to thank the following companies for providing photographs for this volume:

Adel Tool Co.
Armstrong Bros. Tool Co.
Bernzomatic Corp.
Black & Decker Manufacturing Company, Ltd.
Brookstone Company
Dremel Manufacturing Co.
The Foredom Electric Co.
Klein & Sons Inc.
Millers Falls Company
Nicholson File Company
Petersen Manufacturing Company
Rockwell International, Power Tool Division
Skil Corporation
Snap-on Tools of Canada, Ltd.
Specialty Plastics Co.
Stanley Tools
The L. S. Starrett Company
Ungar
J. H. Williams & Co.
Wiss & Sons, Co.
X-acto
Xcelite Incorporated

Contents

Miscellaneous Hand Tools

1-1. General

This chapter deals with miscellaneous hand tools which do not fit into any of the other categories in this book, but which are very useful, if not necessary, for many jobs. Each section is self-contained and includes descriptions of the tool itself, its application, and its care. It is advisable that you read the introductory write-up for each tool in this chapter to determine if you need it in your shop.

The miscellaneous tools covered in this chapter are: *abrasives, awl, glass cutter, knives (modeller's/carver's knives* and *gouges), utility knives, nailset, nibbler, oilstone (sharpening* or *whetstone), propane torch, putty knife (scraper), ripping bar (pinch bar, pry bar), riveter, sanding blocks, scraper (wood, paint* and *glue), terminal (electrical) crimper, tin snips* and *wire stripper.*

1-2. Abrasives

Abrasives are used to prepare wood, metal and other materials for the application of a final finish such as paint, varnish or lacquer; to polish metals, stones, plastics and ceramics to very smooth, bright finishes; to remove paint and rust; and to clean workpieces.

Three kinds of abrasives are available: papers, powders and wool. Abrasives with backings are often referred to as sandpaper, although the abrasive is not sand, nor is the backing always paper. For this discussion, abrasives have been classified into *abrasive papers* and *steel wool.* The only powders mentioned are pumice and rottenstone, which are included in Table 1-2.

1-3. Abrasive Papers

Abrasives adhere to backings of paper, cloth, fiber or plastic, or paper and cloth combined. A paper backing is usually used for hand

Table 1-1. Sizes of Abrasive Papers

Format	Size
Sheets	9 x 11, 4-1/2 x 5-1/2, and 3-2/3 x 9 inches
Belts	2 to 12 inches wide
Cylinders	1-1/2 to 3-1/2 inch diameters
Tapes	1 to 1-1/2 inches wide
Discs	3 to 14 inch diameters

sanding; the other backing materials are used for machine sanding. Abrasives are attached to backing in an *open coat* or a *closed coat* fashion. An open coat backing is 50 to 70 per cent covered with abrasive. While it provides less cutting and more flexibility than the closed coat, the open coat prevents the abrasive from becoming clogged with residue. With a closed coat backing, the abrasive covers 100 per cent of the backing. Self-cleaning papers, which are used with belt sanders to remove glue from surfaces, have

Table 1-2. Abrasives

Abrasive	Color	Description	Use
Silicon carbide	Dark gray	Hard, brittle, sharp.	Cuts soft metals and plastics. Smooths and frosts glass (by rubbing). Used in floor sanding.
Aluminum oxide	Brown	Long-lasting, fast, tough	Used on power sanders for sanding wood and metal. Sharpens tools. Shapes and polishes metal.
Garnet	Reddish brown	Inexpensive, tough	For hand sanding clean wood and general woodworking.
Flint	Yellow-white	Inexpensive, doesn't last long. Paper clogs easily. Soft.	For sanding painted wood or metal. For sanding gummy wood.
Emery	Black	Previously the best abrasive for metal, it is being superseded by silicon carbide and aluminum oxide.	For metal polishing (non-plated metals).
Crocus	Red	Ferric oxide	For polishing soft and non-ferrous metals.
Pumice	Off-white	Powder. Form of volcanic glass. Cuts faster than rottenstone.	For hand-rubbed finishes. Apply with felt pad dampened with linseed oil.
Rottenstone	Off-white	Powder. Decomposed siliceous limestone.	For hand-rubbed finishes. Apply with felt pad dampened with linseed oil.
Steel Wool	Silver gray	Available in several grades.	Scouring, removing paint and rust. Polishing metal.

Table 1-3. Abrasive Papers (excluding emery cloth)

Uses	Number Size	Grit No.	Grade
Rapid removal of surface material, such as sanding old floors, removal of paint and rust.	4 1/2	12	Extra coarse
	4 1/4	14	↑
	4	16	
	3 3/4	18	
	3 1/2	20	
	3 1/4	22	
Removal of surface material, as new floors.	3	24	
	2 1/2	30	↓
	2	36	Extra coarse
Removal of rough stock. Occasionally used for rough wood and paint removal.	1 1/2	40	Coarse
	1	50	↕
	1/2	60	Coarse
Removal of light stock. Use to sand walls before painting	1/0	80	Medium
	2/0	100	Medium
Preparation for finish; use prior to primer or sealer.	3/0	120	Fine
	4/0	150	↕
	5/0	180	Fine
Finish sanding between coats of paint, laquer or varnish.	6/0	220	Extra Fine
	7/0	240	↑
	8/0	280	
	9/0	320	
	10/0	400	
Used rarely for woodworking, but is used for plastics, stone, metals and ceramics.	11/0	500	↓
	12/0	600	Extra fine

soap between the grains to prevent clogging. Backings are also classified as *standard* or *wet;* residue may be washed away from the wet type.

Abrasive backings are also classified by letter designators: A, lightweight; C, heavier for hand sanding; D, heavier for machine sanding; J; and X which is heavier and less flexible than J.

Abrasive papers are available in sheets, belts, tapes, discs, rolls and cylinders. Sizes vary as shown in Table 1-1.

You may tear abrasive papers to fit a sanding block or a sanding machine as follows: fold the paper one way and crease it. Then unfold the crease and fold the paper in the opposite direction. Place the crease over a straight edge and tear.

Clean clogged abrasive paper with a stiff brush, a file card (see Section 6-3 in Volume

Table 1-4. Emery Cloth

Uses	Number Size	Grit No.	Grade
Rust removal and removal of material.	3	36	Coarse
	2 1/2	40	↕
	2	50	Coarse
Cleaning work tools and rust removal.	1 1/2	60	Medium
	1	80	↕
	1/2	100	Medium
Hand polishing of non-plated metals.	1/0	120	Fine
	2/0	150	↕
	3/0	180	Fine

Table 1-5. Steel Wool

Uses	Number Size	Grade
Paint, rust, soot and dirt removal, often used with paint remover.	3	Coarse
Floor maintenance, removing soiled wax, and preparation of surfaces for refinishing.	2	Medium Coarse
For kitchen use and for cleaning whitewall tires. Cleans and polishes aluminum and brass, revives surfaces, and is used for floor buffing. Removes tile stains (don't use on plastic tile).	1 0	Medium Fine
For very smooth finish and for rubbing down final coat. Also for cleaning brass, aluminum and copper before soldering.	00	Very fine
For extra-smooth finish and for rubbing down final coat of a finish.	000 & 0000	Extra fine

3) or a wire brush. A light tapping will also remove some of the residue.

Table 1-2 lists the abrasives most commonly used by home craftsmen, hobbyists and do-it-yourselfers. Except for pumice and rottenstone, which are powders, all the abrasives listed adhere to backings.

Table 1-3 lists the grades, grit number, number size and uses of the various grades of abrasive papers. The grit number is the number of openings in a screen through which abrasives can pass. The openings vary from 12 (very coarse grits) to 600 (very fine grits).

Emery cloth is used dry or with oil primarily for light cleaning of tools, for rust removal, and for hand-polishing non-plated metal surfaces. Emery cloth is made of grits of the abrasive mineral emery on a backing similar to other abrasive papers. Emery is actually a natural mixture of aluminum oxide and magnetite. Emery cloth is available in 9″ × 11″ sheets and grades as listed in Table 1-4. Emery cloth, once the best abrasive for use on metal, is now being superseded by silicon carbide and aluminum oxide.

1-4. Steel Wool

Steel wool is an abrasive material composed of long, fine, steel shavings. It is used especially for scouring, removing rust from tools and polishing metal. Table 1-5 describes the grades and uses of steel wool.

1-5. Awl

The awl (Figure 1-1) is a handy, inexpensive tool to have in your toolbox. Shaped like an ice pick, the awl is used to make starter holes for small brads, as when hanging pictures, to make holes in leather and similar materials, to make pilot holes in wood for starting a drill bit, and to scribe wood. The awl should not be used to scribe metal or to make small holes in metal.

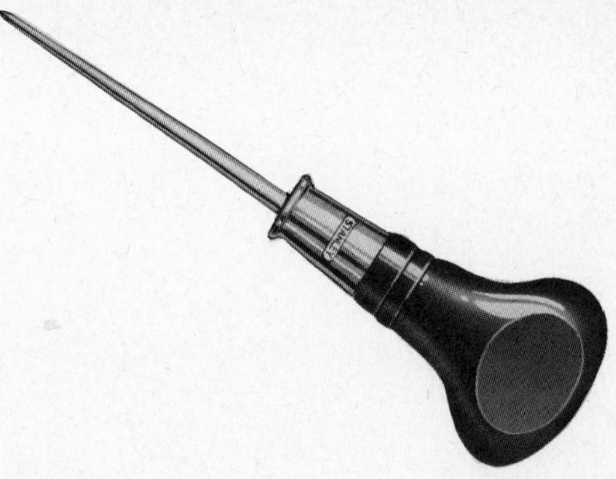

Fig. 1-1. The awl is used to make starting holes for brads, drill bits, and holes in leather or similar materials.

The awl has a wooden or plastic gripping handle and steel spike; its overall length is about 5″. Keep the point sharp by occasionally honing the tip. If the point bends or becomes too dull, sharpen it on a grinding wheel to a long, conical point, and then hone. Don't let the tip become excessively hot during the grinding.

In using the awl to make a pilot hole, push the awl far enough into the workpiece to enable the next tool (or fastener) to be accurately set into the hole. Note that the awl is not a drill; it pushes the workpiece fibers apart rather than cutting them. On thin, small workpieces, this separating of fibers can split your workpiece, so don't make your pilot hole any larger than necessary. In precision drilling of a workpiece, the correct procedure would be to make a point with the awl at the starting mark. Then, drill a hole with a small drill such as a 1/16″ or 5/64″ drill; continue with increasingly larger drills until the desired diameter is reached.

1-6. Glass Cutter

The glass cutter consists of a malleable iron handle with a ball on one end and either a tungsten steel or a tungsten carbide wheel on the other end. The tungsten carbide wheel is much more expensive than the tungsten steel wheel and is needed only if you intend to do a lot of glass cutting. The wheel is used to score a line along the surface of the glass. The ball is used in some instances to tap the glass, causing it to separate at the scored line. Notches under the wheel are also used in some instances to break or nibble small pieces of glass from the workpiece.

WARNING: Protective gloves should be worn while cutting glass.

The following steps tell how to cut glass:

1. Cover a large flat surface with layers of newspapers. Clean the glass to be cut and place it on the newspapers.

2. Wipe the glass along the proposed cut line with a light oil or turpentine. Apply a drop of light oil to the glass cutter wheel.

3. Hold the glass cutter vertical with the notches toward the glass. Your forefinger is placed on the indentation in the handle.

4. Place the glass cutter along a straight edge of a piece of scrap wood that is used as a guide.

5. With continuous firm pressure, draw the glass cutter wheel along the glass so that the glass is *scored* - do not attempt to cut through the glass. Instead, strive for one continuous line.

6. Place a pencil or a dowel *underneath* the glass and along the scored line. Immediately press with your hands on each side of the scored line (or press firmly on one side and slap the other side with the open

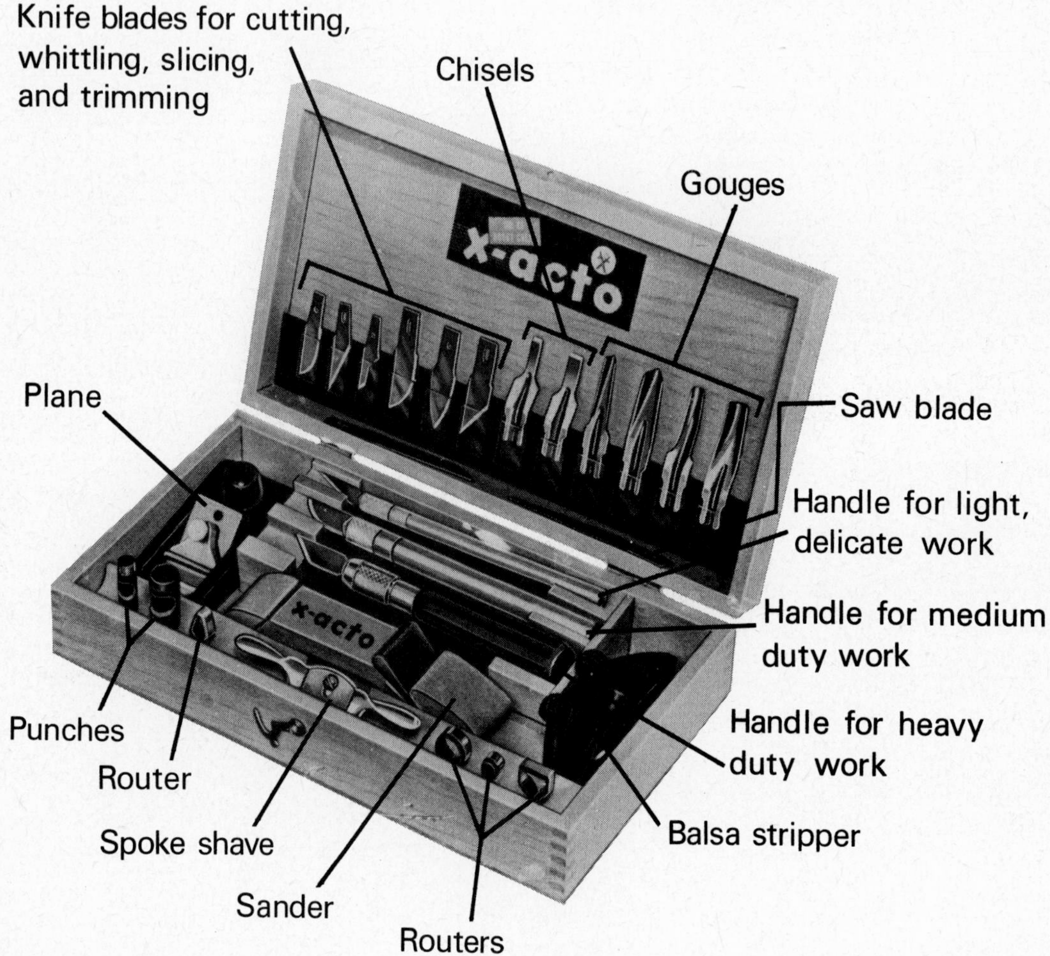

Fig. 1-2. This modeller's/ carver's workshop contains knives, chisels and gouges as well as other handy tools.

palm of your other hand). You may also place the glass over the edge of a table; hold the glass with both hands and snap down on the side which lies over the edge.

To cut narrow strips, score the glass. Then use the ball on the glass cutter handle and tap the glass *underneath* the scored line. For cutting off very narrow strips, score the glass. Then use pliers (duck bill preferred, Section 3-8) to break (nibble) the glass off. You can also use the notches on the glass cutter to break the glass off.

Some additional hints for successful glass cutting are: never tap on the scored side of the glass; retrace with the cutter wheel only over spots where the cutter missed since retracing over a scored line will result in chipped, rough edges; practise on scraps of glass before cutting the workpiece; use moderate pressure on the cutting wheel – too much pressure causes chipping and too little results in unscored spots along the cutting line.

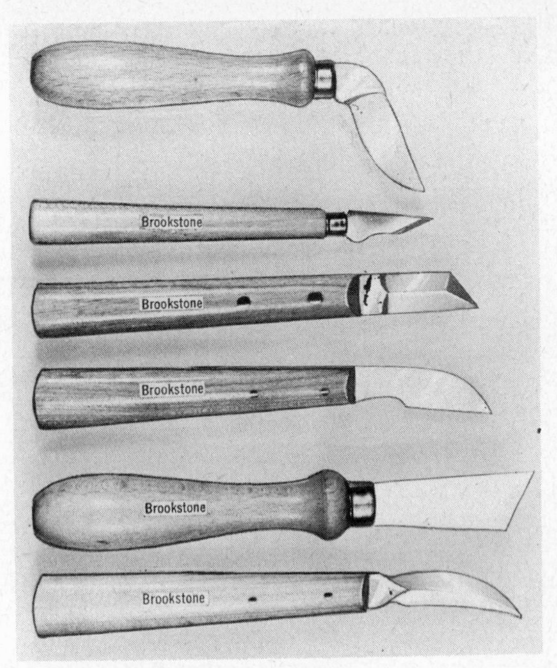

Fig. 1-3. Wood-carving knives.

1-7. Knives (Modeller's/ Carver's Knives, Chisels, Gouges)

Modeller's and carver's knives, chisels and gouges are used to carve, sculpture, cut patterns, make models, and make and repair furniture. The workpiece materials may be wood, soap, linoleum or other similar materials. The tools are available in sets (Figures 1-2 and 1-3) or as individual pieces.

The knives are used to chip, carve and whittle, and come in many varied shapes, sizes and lengths of cutting edges. The chisels have flat blades and make a straight, stabbing cut into the workpiece. The gouges have curved blades and make curved, stabbing cuts. One exception to this is the V-shaped or U-shaped tool which is sometimes known as a *veiner.*

Accessories which the modeller/carver may want include routers, punches, sanding blocks (Section 1-16), spoke shaves (Section 2-6), planes (Chapter 2), a balsa stripper, C-clamps (Section 4-7, Volume 3), oilstones (Section 1-11), and a wooden mallet (Section 8-6, Volume 3). Routers carve grooves or hollow the interiors of models. Punches are used to cut small pieces of round dowel and the balsa stripper is used to guide a knife in cutting thin strips of balsa wood. The wooden mallet is used to strike the handle of the gouges and chisels when deep cuts are made.

There are two basic methods to cut and carve with knives. One method is the thumb-push method. The workpiece is held in the left hand, the knife is held in the right with the thumb on the back of the blade. The blade is pointed away from the body so that the cut is made in that direction. The left thumb is placed on top of the right thumb and pushes the knife.

The second method is the draw method. The workpiece is held in the left hand and the knife in the right. The cutting is done toward the body with the left hand supplying the power against the right fingers that are guiding the knife. The right thumb steadies the workpiece.

Workpieces are first rough-cut to shape

with a coping saw (Section 6-7). Various knives, gouges, and chisels are then used to shape the workpiece. Sanding may be performed to complete the workpiece.

When buying chisels, gouges and knives, be sure to obtain tools made from good-grade tempered steel. They should have long, comfortable handles, and they must be sharp to ensure easy, accurate cutting. New carving knives, chisels and gouges need honing.

Sharpen your carving tools on an oilstone – use a flat oilstone for blades and chisels, and an oilstone with convex and concave surfaces for sharpening gouges. Place a few drops of oil on the oilstone before using it. Place the tool bevel against the oilstone at the same angle as the manufactured bevel. Slide the tool across the stone to sharpen the edge. When the bevel is smooth and a slight burr is raised, lift the tool from the oilstone. Rub the oilstone over the burr very carefully until the burr is removed. The edge of the tool can then be stropped by carefully drawing the edge across a piece of leather.

Knives are sharpened by alternately pushing and pulling the blade edge over an oilstone in a circular motion. Test for burrs with your finger and remove them with an oilstone. Finally, strop the edge on a piece of leather.

Protect knives, gouges and chisels from dropping or bumping each other or other tools during storage. Wipe the steel surfaces with an oily rag occasionally to prevent rusting.

1-8. Utility Knives

Two utility knives which are handy around the home and shop are shown in Figure 1-4. The lineman's or electrician's knife has three blades made of high-grade tempered steel. The wire-skinning blade is 1″, the spearpoint general-purpose blade is 2-1/2″, and the screwdriver blade, which locks open, is 3″.

The retractable blade knife (6″) is an aluminum handle with a latching device to hold a steel razor blade into one of three cutting positions. Blades are replaceable

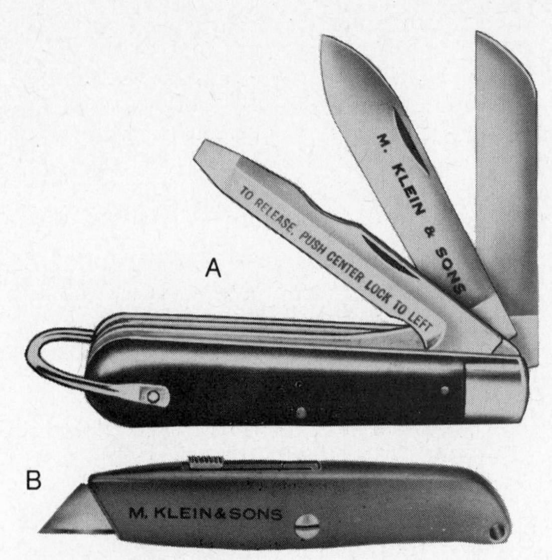

Fig. 1-4. Utility knives: (A) lineman's or electrician's knife; (B) retractable blade.

(spare blades are stored in the handle). This knife is particularly useful in cutting asbestos shingles, tar paper, underlay paper, and asbestos vinyl and vinyl tiles. It is also very handy for trimming wallpaper.

To sharpen a pocket utility knife, hold the blade against a whetstone at a 30° angle. Stroke the knife forward diagonally, then turn it over and return the stroke, making sure the entire blade is sharpened. Sharpen the edges alternately on the stone, and then strop on a piece of leather.

1-9. Nailset

It is unsightly to have nailheads showing on the surface of a piece of furniture or other workpiece. To prevent nailheads from showing, craftsmen use finishing nails (Section 5-8, Volume 3) and countersink the heads from 1/16″ to 1/8″ below the surface with a nailset (Figure 1-5). The remaining hole is then filled with a wood filler or putty and the workpiece is finished.

A high-quality nailset is made from an alloy steel and is hardened and tempered for long life. It has a bevelled, square head that

Fig. 1-5. The nailset is used to countersink nailheads below the surface of the workpiece.

prevents mushrooming and rolling on a slanted surface. A knurled handle prevents the nailset from slipping in your hand. The nailset point is cupped so it won't slip off the nailhead and mar the surface. Nailset point diameters are marked on the head and sets are available with diameters of 1/32″, 1/16″, 3/32″, 1/8″, and 5/32″. The body length is 4″ with an 11/32″ body diameter.

To countersink a nail, first drive the finishing nail with a claw hammer to about 1/16″ from the surface of the workpiece. Select the proper nailset for the nail diameter being driven. Place the nailset into the nailhead and position the nailset so that it is in line with the nail. Tap the nailset head with the hammer; before tapping again, make sure that the nailset point still remains in the nailhead. Drive the nail 1/16″ to 1/8″ below the surface.

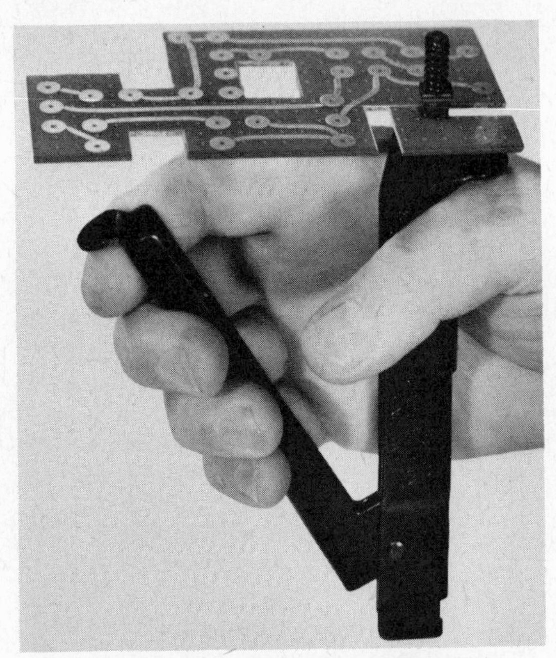

Fig. 1-6. The hand nibbler is used to cut holes, slots, or other openings in plastic, vector board, and metals.

1-10. Nibbler

The hand nibbler (Figures 1-6 and 1-7) is used to nibble holes, slots or openings of any shape in steel up to 18 gauge (0.046″), and in plastic, vector board, copper, aluminum or other soft metal up to 1/16″. The cut edges are safe, flat and smooth. The nibbler, a rectangle approximately 1/16″ by 1/4″, is useful in duct work, auto body work, and in making chassis for electronic units. The unit shown in Figure 1-7 can be used to cut tubing lengthwise.

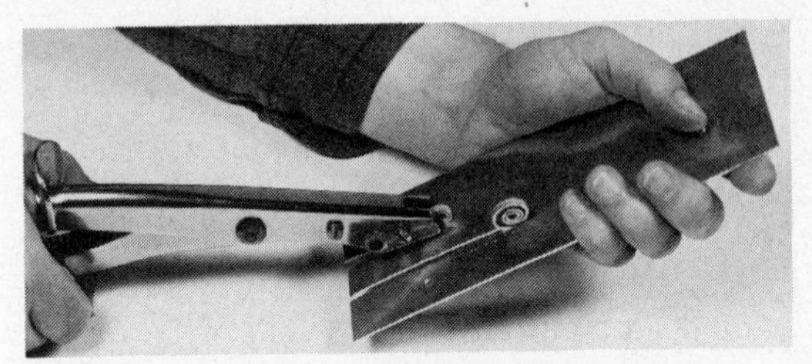

Fig. 1-7. This hand nibbler is being used to cut slots in a workpiece.

To use the nibbler, first cut a 1/4″ square hole or drill a 1/4″ hole in the workpiece to allow the nibbler punch to pass through. Insert the nibbler punch and squeeze the handles together and follow the layout pattern around the workpiece. Make the nibbler bites as close together as possible. File any rough edges.

When the nibbler punch becomes dull, it may be replaced with a new punch. Remove and replace the punch in accordance with the manufacturer's directions.

1-11. Oilstone (Sharpening Stone or Whetstone)

In addition to being difficult to use, dull tools do a poor job and are dangerous because they require more force than sharp tools to make them work. You should use an oilstone to maintain the sharpness of such tools as knives, chisels, gouges, scraper blades, awl points, plane blades, spoke shave blades. Use it also to remove small burrs from the edges of workpieces.

Oilstones are made either of vitrified aluminum oxide (best for all-round use) or vitrified silicon carbide. One side of the oilstone is coarse and the other side is fine. They are rectangular, tapered, flat, oval and round, and range from pocket size (3″ × 7/8″ × 1/4″) to large size (about 9″ × 1-1/2″ × 1/2″).

New oilstones should be saturated with oil, and kept in a dust-free place such as a wooden box. During use, keep the oilstone lubricated with either a light machine oil or equal parts of machine oil and kerosene.

Loose grit and sludge should be removed from the surface of the oilstone by rubbing it with a solvent-soaked rag. Gummy residue can be removed by warming the oilstone in the oven and then wiping off the oily sludge as it comes to the surface. The oilstone can be ground flat again by rubbing it in a mixture of silicone carbide grit and water that is placed on the surface of a piece of glass.

Procedures for using the oilstone to sharpen specific tools are described separately in the appropriate section of this book. In sharpening any tool, remember that sharp tools reflect light. When a tool is held up to light, a white blurred line means that the tool is not sharp. In honing, move the tool over the complete surface of the oilstone so that it wears evenly.

1-12. Propane Torch

You can find many uses for a propane torch around the home, shop or farm: removing paint and putty, soldering copper pipes, wires and gutters, unfreezing pipes, removing tile, refinishing furniture, lighting charcoal fires and soldering art objects.

The propane torch comes in a set (Figure 1-8) and consists of a propane cylinder (about 15 hours' use), control valve, blowtorch head, flame spreader, solder tip, pencil flame head and a flint spark lighter. The

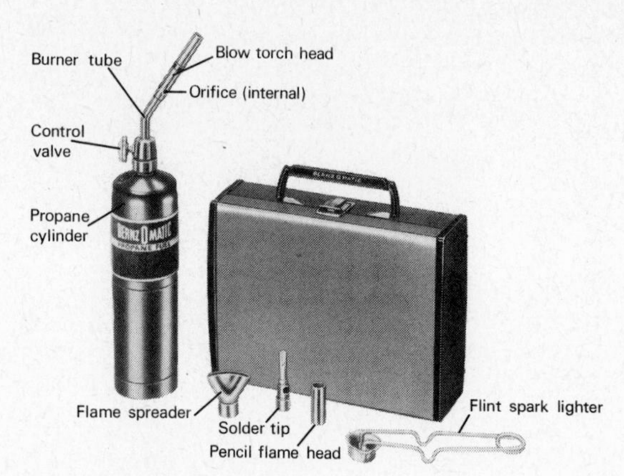

Fig. 1-8. The propane torch.

cylinder contains propane, which is a liquid under pressure in the cylinder. When the control valve is opened, propane escapes and expands to 400 times its size in volume. The gas is combined with air to produce an intense flame. The flame spreader spreads the flame so heat is distributed over a large area. The solder tip is heated internally by the flame and is then used to solder wires or other suitable metal parts. The blowtorch head and the pencil flame head control the air mixture and hence, the intensity of heat coming from the torch. The flint spark lighter is used to light the torch.

NOTE: The flame spreader is always used on the blowtorch head; the soldering tip is always used on the pencil flame head.

The propane torch is easily assembled. First determine the head to be used and select the correct orifice to use with it: for a pencil flame head, use brass 0.005 orifice; for a blowtorch head, use nickel 0.008 orifice. Insert the correct orifice into the burner tube. Screw the head onto the burner tube hand-tight; use a wrench to turn the head just slightly tighter. Add the flame spreader or solder tip to the head, if desired; tighten the attachment screw hand-tight. Make sure the control valve is off by turning it clockwise as far as it will go. Place the valve unit to the top of the replaceable propane cylinder, insert, push in and turn simultaneously until the burner unit is secured hand-tight to the cylinder. Do not tighten with a wrench.

CAUTION: Propane is highly flammable. Keep the propane cylinder safely stored

Table 1-6. Solders and Alloys

Type of solder or alloy	Stainless steel solder and flux	General-purpose acid core solder	All-purpose resin core electrical solder	Alu-minum brazing alloy and flux	Silver solder and flux
Apply solder or alloy when:	Solder flows freely on contact with heated metal	Solder flows freely on contact with heated metal	Solder flows freely on contact with heated metal	Flux becomes a clear liquid	Flux becomes a thin, clear liquid and forms dull red
For metals below use solders or alloys marked "X" at right					
Aluminum For strength in joining sheets, sections, etc.				X	
Chrome Plate For trim, when on steel, brass, copper or nickel alloys. (Not on die castings.)	X				
Copper or Bronze For electrical equipment, or fittings, tubing, utensils, etc.		X	X		
Galvanized Iron or Steel For cans, buckets, tanks, eaves-troughs, etc.		X			
Silver and Silver Plate For jewelry, flatware, etc.	X				X
Stainless Steel For appliances, kitchen equipment, or wherever strength is needed	X				X
Steel For utensils, pipes, sheets, tool steels, motors, etc.	X				X
Unlike Metals Such as steel to brass	Unlike metals with X's in the same vertical column can be joined. For example: Copper and galvanized iron, with general-purpose solder.				
Dressing stone	Dress emery points.				
Mandrels	Used to mount polishing accessories, cutting wheels, sanding discs, and polishing wheels.				

away from heat and flame. Discard the empty cylinder in a safe place – do not throw it in a fire. Keep the cylinders out of the reach of children. Remove the burner from the cylinder when the torch is not in use.

To light the torch, open the control valve (turn counterclockwise) until a low hiss of escaping gas is heard. (If the blowtorch head is being used, open the valve one-half turn further.) Immediately ignite the torch with a spark from the flint spark lighter or from the flame of a match. Let the burner remain on a small flame until it is hot before opening the control valve any further.

Adjust the control valve for flame size. For maximum efficiency, adjust the flame from the pencil flame head so that the blue cone at the center of the flame is 1-1/2″ long; for the blowtorch head, the cone should be about 5/8″ long.

To solder with the propane torch, first tin the soldering tip in accordance with the procedures described in Section 4-13. Use a small flame against the tip at all times and re-tin as necessary. Thoroughly clean the surface to be soldered with sandpaper, emery or steel wool. Apply the proper soldering flux. Preheat the surface to be soldered with a flat face of the soldering tip. Apply the solder to the surfaces to be soldered, not to the soldering tip. Let the solder flow. Do not move the workpiece until the soldered area is cool.

You can use the flame spreader attachment on the blowtorch head for removing cracked or broken tiles. Warm the tiles sufficiently with the torch to soften the adhesive underneath. Using a stiff-bladed knife, putty knife, or a pick, remove the tile. Again apply heat to the adhesive to warm it. Put the new tile in place and press it firmly.

To thaw out pipes, open the faucet that does not deliver water. Starting at the faucet, work slowly with the flame until a free flow of water is restored. Do not start heating the pipe at some intermediate point. Do not overheat soldered pipe joints, nearby studs or asbestos board behind the pipe. The latter act as heat shields.

Remove putty and paint as follows: using a low flame, apply heat slowly along the putty to soften it. Use a scraper to remove it. Use the flame spreader attachment on the blowtorch head to soften paint for easy removal. Remove the paint with a scraper as you move the flame obliquely over the surface (never at a right angle to the surface).

The torch can also be used to heat conduit for easy bending. Secure the conduit and place a piece of BX cable through it to prevent the conduit from creasing when it is bent. Place a piece of pipe over one end to aid in bending the conduit. Heat the conduit for several minutes with the torch by moving the flame slowly back and forth across the area to be bent. Bend the conduit, quench the heated section with water, and withdraw the BX cable.

If the flame is inefficient or will not light, the orifice is clogged and must be cleaned. Using a wrench, remove the head from the burner tube: be careful to check the direction of the orifice when it comes out of the burner tube. Now reverse the direction of the orifice, place it in the head, and screw the head hand-tight back onto the burner tube. Open and close the control valve rapidly several times to blow gas and impurities through the orifice. Remove the head, reverse the orifice, and reassemble the head to the burner tube. Tighten hand-tight and then use a wrench to turn the head just slightly further.

Oxidation and charring may be removed from the heads and accessories with steel wool. Store the torch in a safe place away from heat, flames and children.

1-13. Putty Knife (Scraper)

A putty knife or scraper is a tool of many uses. It can be used to remove paint, old gaskets, wallpaper, undercoating and grease; to apply plaster, wood filler, putty and auto body filler; to scrape furniture. In a pinch, you can use a stiff-bladed putty knife as a prying tool to open paint cans, to screw and to chisel.

High-quality putty knives (Figure 1-9) have

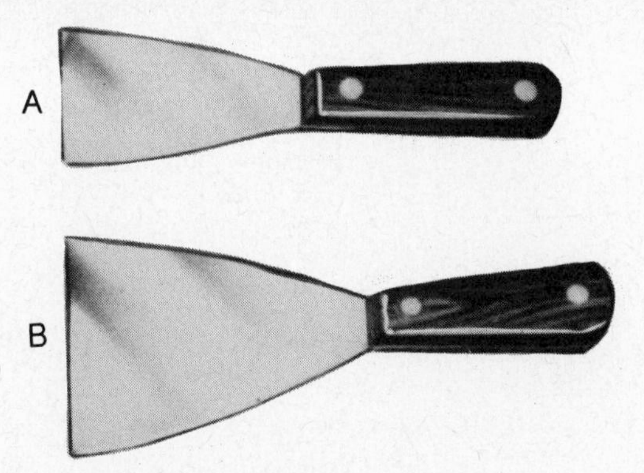

Fig. 1-9. (A) Putty knife and (B) scraper.

mirror-finished, high-carbon steel blades that are hardened, tempered and ground. Blades should pass completely through shatterproof handles. Some blades are straight while others have chisel edges. Knives are available with either stiff blades that are useful for scraping or flexible blades that are useful in applying fillers and putties and in removing wallpaper.

Narrow-bladed tools are classified as putty knives; wide-bladed tools are classified as scrapers. Blade widths range from 1″ to 4″.

Keep putty knives and scrapers clean. Sharpen edges with an oilstone or by regrinding if necessary.

1-14. Ripping Bar (Pinch Bar, Pry Bar)

Ripping bars, pinch bars and pry bars all have one function: to pry two objects, such as nailed boards, crates, or stuck windows, apart. The only differences between the various bars are in the types of ends and the sizes (Figure 1-10).

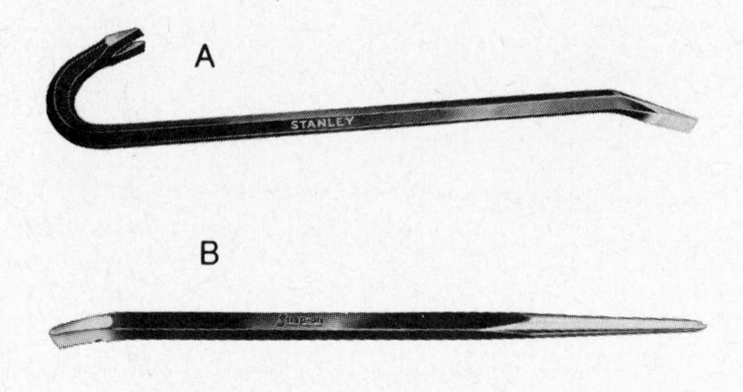

Fig. 1-10. (A) Ripping bar and (B) pinch bar.

Pry bars, the smallest, are between 7/8″ × 7″ and 1-3/4″ × 14-1/2″ with one or two ends clawed for nail pullings. Pinch bars are from 1/2″ × 16″ to 7/8″ × 42″ and have a long, tapered end with a sturdy line-up tool on one end and a blunt chisel tip on the other end. The ripping bars are from 1/2″ × 12″ to 7/8″ × 36″ and have one chisel end and one slotted claw (see ripping chisel, Section 3-6, Volume 3).

To pry two objects apart, start first with the smaller pry bar and work up to the larger pinch or ripping bar as required. Place the clawed edge under the board or nail to be pried loose. Press on the bar. The tool leverage can be increased by placing a piece of scrap wood under the offset bend. Wedges can be placed between the partially separated pieces.

1-15. Riveter

Rivets are used to join, among other things, two workpieces made of thin metal, wood, canvas, leather, plastic or fiber. They may be used in place of screws, bolts, nuts, spot welds and solder joints, and are easy to use, fast, effective and economical.

The rivet tool shown in Figure 1-11 installs rivets from one side only – this operation is called *blind riveting* because the user doesn't need to see the other side of the material. This feature makes this riveter very popular for home, farm and industrial use. The rivets used in this type of riveter have clinching mandrels through the center of the rivets. The rivet tool has two handles that cause a pair of jaws to grip the rivet mandrel and pull the mandrel through the rivet head, flattening the head against the workpiece. The riveter is about 10″ long.

Figure 1-12 illustrates two workpieces being joined together by a blind rivet installa-

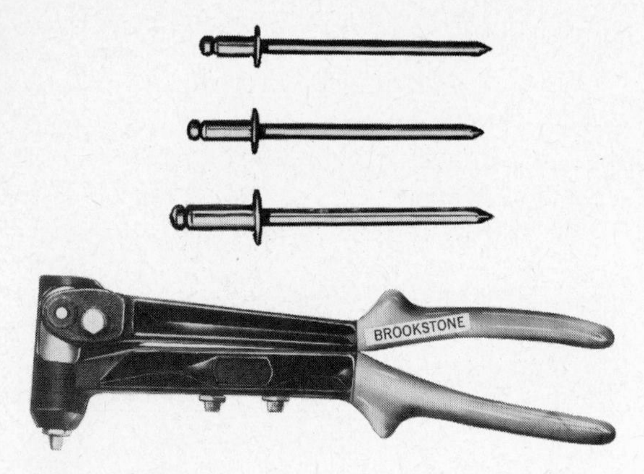

Fig. 1-11. Riveter.

tion. As the top of the illustration shows, a rivet with a clinching mandrel is inserted through a predrilled hole. The riveter jaws grasp the mandrel and pull the head of the mandrel against the rivet, which in turn flattens the rivet against the workpiece and makes a tight joint. When the rivet and joint are tight, the mandrel breaks and falls free, leaving a neat riveted joint on both sides of the workpieces.

Rivets for the rivet tool come in several diameters, including 3/32″, 1/8″, 5/32″ and 3/16″. Rivet lengths vary also: use 3/16″ long rivets for thicknesses up to 1/8″, 1/4″ for up to 3/16″ thickness, 3/8″ for up to 5/16″ and 7/16″ for thicknesses up to 3/8″. If you are using very thin workpieces, you can add a flat washer onto the rivet to give extra thickness. The proper rivet should protrude no more than one rivet diameter through the thickness of the materials.

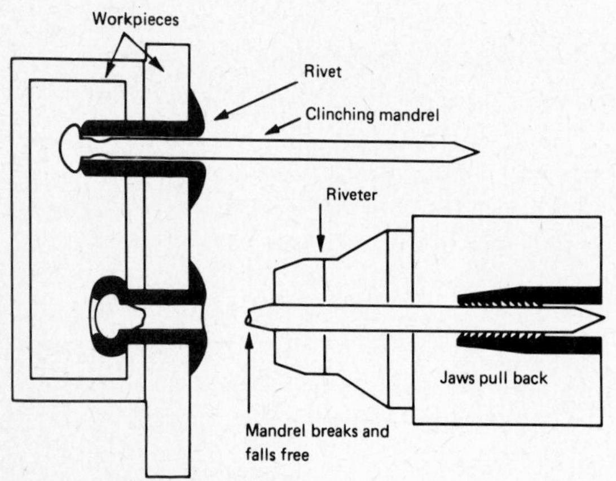

Fig. 1-12. Blind rivets.

To use the riveter, first select the rivets to be used. Predrill holes of the same diameter as the rivets to be used. Check that the rivet fits through the workpiece properly. Open the riveter handles. Select the proper hole diameter on the riveter head and screw or slide (as applicable) the proper tool head into position. Place the rivet into the tool and close the handles slightly to grip the rivet. Insert the rivet into the predrilled hole and squeeze the handles. Alternately open and close the riveter handles until the mandrel breaks off. Shake the mandrel free from the riveter.

1-16. Sanding Blocks

Sanding blocks hold abrasive papers (Section 1-3) and provide a flat surface that distributes even pressure over the workpiece. The blocks not only keep soft areas in wood workpieces from being sanded deeper than the harder fibrous areas, but also prevent rounding of the edges and corners of the workpieces.

You can buy a sanding block similar to the one in Figure 1-13 or you can make your own to suit your need. The plastic body block shown fits into your hand comfortably. A 2″ × 7″ rubber pad under the abrasive paper facilitates sanding a workpiece to a smooth, flat surface. This model even dispenses its own sandpaper from a five-foot roll

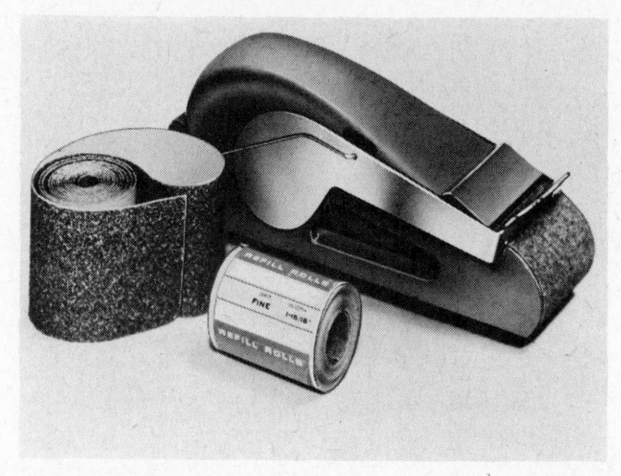

Fig. 1-13. Sanding blocks may be bought or made in the shop.

stored inside. To change the abrasive paper, lift the release lever, tear off the used paper, pull out a new length and close the lever.

Sanding blocks with felt pads and a metallic bottom with tungsten carbide chips affixed to the metal are available. These blocks are approximately 5-1/2″ × 3″ and are very durable. Residue can be brushed from the grits with a file brush; gummy residue can be removed with a solvent.

If you make your own sanding blocks, a convenient size is 3/4″ × 5-1/2″ × 3-1/4″. Standard 9″ × 11″ sheets of abrasive paper can be torn into four equal pieces. The paper can be held on by hand or may be stapled or tacked into place along the edges. For special applications, sanding blocks can be designed with curved or rounded edges and in various sizes.

When sanding wood with the sanding block, stroke with the wood grain unless you are doing very rough sanding, in which case it is all right to sand across the grain. Sanding across the grain leaves cuts that are sometimes difficult to remove. Sand from one end of the workpiece to the other in continuous strokes. Use coarse-grit abrasive papers first; then use fine-grit papers until you attain the smoothness you desire.

1-17. Scraper (Wood Scraper, Paint and Glue Scraper)

The scraper (Figure 1-14) is used to scrape wood, remove paint, glue and barnicles, and to clean grease or other residue from surfaces. Scrapers are available with hardwood or aluminum handles in a range of types and sizes from 6″ long with cutter widths of 1-1/2″ to 10″ long with cutter widths of 2-1/2″; the smaller scrapers are handy for window sashes, windows and close-quarter work.

Scraper blades come in many shapes. Flat blades, hooked blades, serrated blades, two-sided blades and four-sided blades are some of the types available. Many have reversible

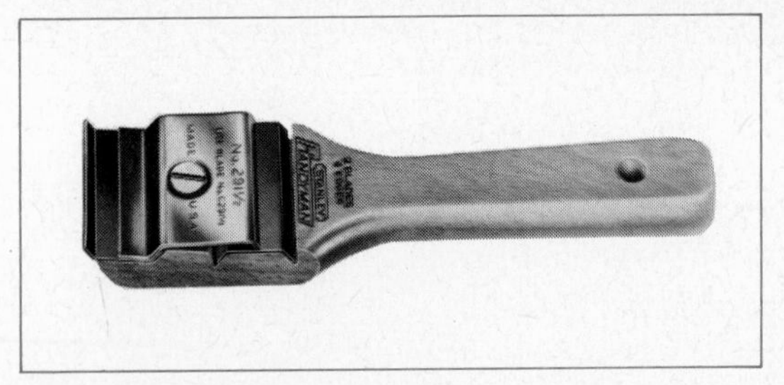

Fig. 1-14. The scraper is used to scrape wood and to remove paint and glue.

blades (the scraper illustrated has a reversible hooked blade and also stores one blade).

Blades may be held in with a screw or by means of two blades driven back-to-back into a slot in the handle. To remove the back-to-back blades, place the handle in a vise. Using an old screwdriver and a light hammer, tap the blades out of the slot. Replace them in pairs.

Hand scrapers without handles are also available. A typical hand scraper is 3″ × 5″.

In using a scraper, hold it flat against the surface to be scraped. Pull with pressure distributed evenly across the blade; don't put more pressure on one side of the blade than on the other because you'll gouge the workpiece.

Sharpen a scraper blade with a file, filing at the same angle as that of the original bevel. Each stroke of the file should pass all the way across the blade. Remove the burr from the back edge of the blade with an oilstone.

1-18. Terminal (Electrical) Crimper

A terminal crimper (Figure 1-15) is used to crimp insulated solderless terminals to ends of wire and to make wire splices. The solderless terminals are handy for making outdoor wiring connections, such as in automobiles where power for soldering is not available. They are also very handy for making connections indoors to television antenna wires, antenna rotator wires, and in wiring loudspeakers.

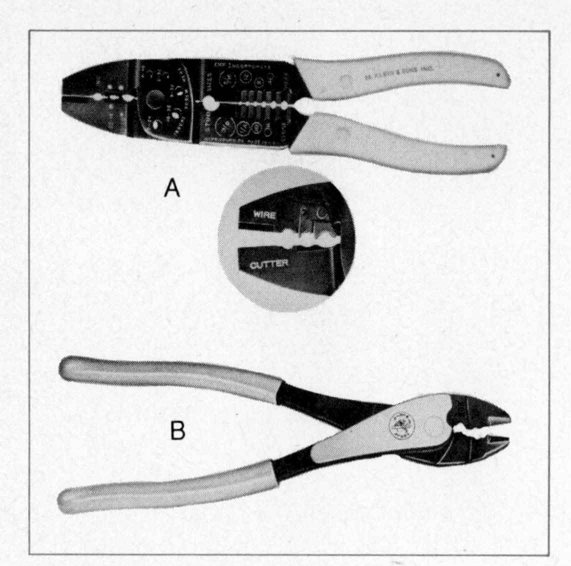

Fig. 1-15. Terminal crimper.

The 8″ terminal crimper shown in Figure 1-15a is an all-purpose tool. In addition to crimping solderless terminals onto no. 22 to no. 10 AWG (American Wire Guide) wire, it also strips no. 22 to no. 10 AWG wire, cuts solid or stranded wire, cuts 4-40,6-32, 8-32, 10-24 and 10-32 screws cleanly without burrs or thread damage, measures wire gauges, and performs electrical and mechanical crimps.

To cut wire, place the wire between the cutting edges and squeeze the crimper handles until the wire is cut. To strip wire, measure the amount of insulation to be stripped off by using the stripping gauge on the side of the crimper. Place the insulated wire into the proper stripping slot. Close the crimper handles; the stripper slots cut through the insulation, but do not cut the wire. With the handles closed, pull the wire so that the insulation is stripped off. Place the terminal in the proper crimper grooved jaws for the gauge of wire being used. The crimper jaws are located at the top of the crimper tool.

To cut a screw or bolt, thread the fastener into the appropriate threads of the tool. The screw length exposed plus 1/8″ (the thickness of one jaw of the tool) will be the length of the cut screw. Squeeze the handles together to cut the screw.

1-19. Tin Snips

Tin snips are hand tools that provide a shearing force for cutting various types of sheet metal, stainless steel and Monel metal up to 18 gauge into various shaped pieces. They are ideal for cutting rain gutter, spout, vent, and ducting installations and for cutting other sheet metals. Forged, high-carbon, heat-treated steel jaws cut smoothly and evenly with a minimum effort. Tin snips are 7″ to 12″ long; jaws cut from 1-3/4″ to 3″ (in one closing of the jaws).

Tin snips are divided into four types (Figure 1-16) according to the pattern cut: regular, straight or curved cuts, left cuts and right cuts. Regular cut tin snips are used for cutting straight lines and curves in locations that are easily accessible. Straight or curved cut tin snips cut straight lines, circles, squares or any pattern; this tin snip is often called a *compound action combination snip* and it's the most useful tin snip for overall metal work. Left cut tin snips cut short straight lines or left cuts in locations where it is advantageous to keep the handles and the hand away from the metal stock. Right cut tin snips cut to the right.

Compound action pipe and duct snips (Figure 1-17) are used to cut sheet metal pipes and ducts, vinyl asbestos tile and asbestos shingles. This snip cuts a 3/16″ waste strip of metal that curls out of the top of the snips; this prevents injury to the operator's hands. Pipe and duct snips can make their own starting hole.

To cut straight lines with tin snips, lay the workpiece on a bench with the guideline extending off the bench. Hold the workpiece with one hand and cut with the other, always holding the tin snips at a right angle to the workpiece. The top blade of the tin snips should cut along the scribed line. The portion of the workpiece being cut off will bend below and out of the way of the tin snips. Keep the tin snips as far into the workpiece as possible so that the cutting is done from a point as near the pivot as possible. Use even pressure. Keep the faces of the tin snip jaws perpendicular to the workpiece to prevent the cut edge

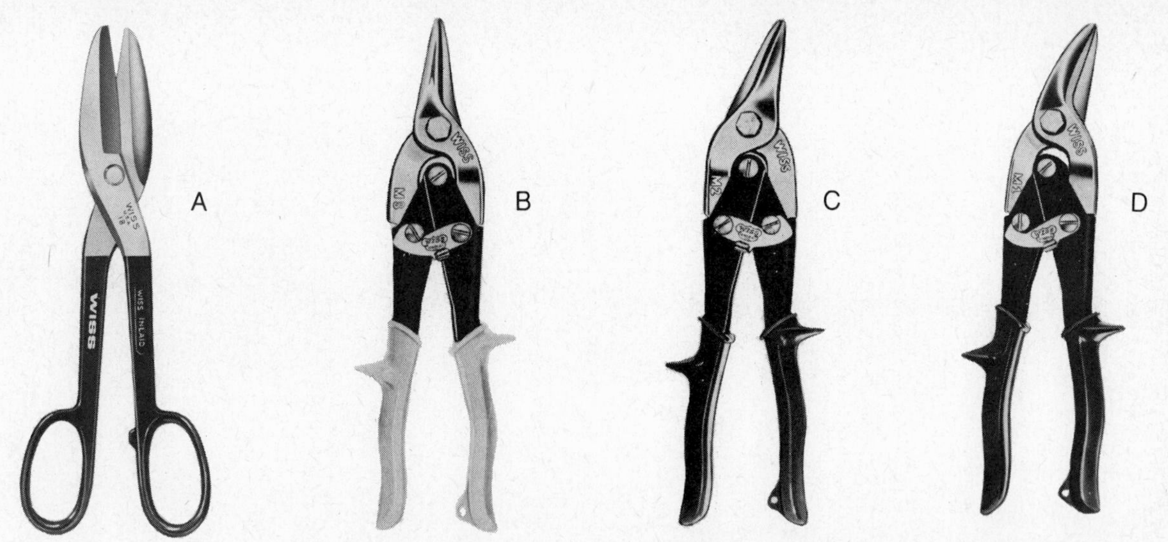

Fig. 1-16. Tin snips: (A) regular; (B) straight or curved cuts; (C) left cuts; (D) right cuts.

from becoming bent or burred. Gloves should be worn to protect the hands from sharp edges and burrs.

When cutting curves, make the cut as continuous as possible. Stopping and starting at different points causes rough edges with sharp slivers of metal. When extra leverage is necessary, place the tin snips on the workbench and bear down on the upper handle.

When a hole or opening is to be cut into a workpiece with tin snips, first cut a hole into the workpiece with a small cold chisel. Make the hole large enough to allow the tin snip blades to enter, open and cut.

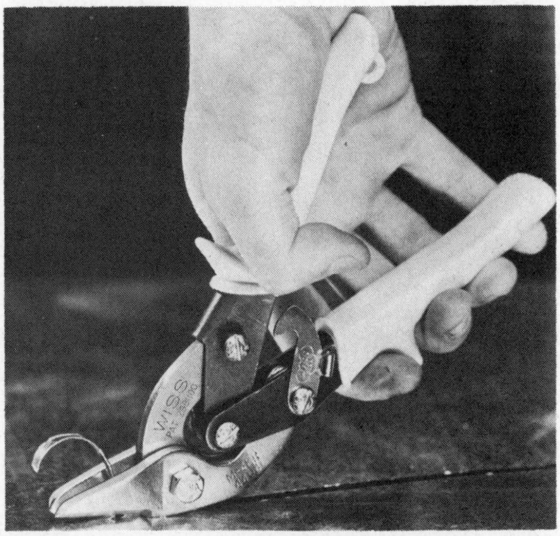

Fig. 1-17. Pipe and duct snips remove a 3/16-inch waste strip from the workpiece.

Do not attempt to cut material heavier than the tin snips can handle. Don't use the tin snips as a hammer.

Dull tin snips can be sharpened with an oilstone or a file. To sharpen, clamp the snips in a vise and use a file as shown in Figure 1-18. The file is used on the bevelled edges only (never on the flat face). Stroke only in the direction shown across the full length of the edge and away from it. To remove the wire edge on the face of the snips (made from filing on the bevelled edge side), draw the face lightly across an oilstone from the pivot to the tip. After use, apply a light coat of machine oil to the snips.

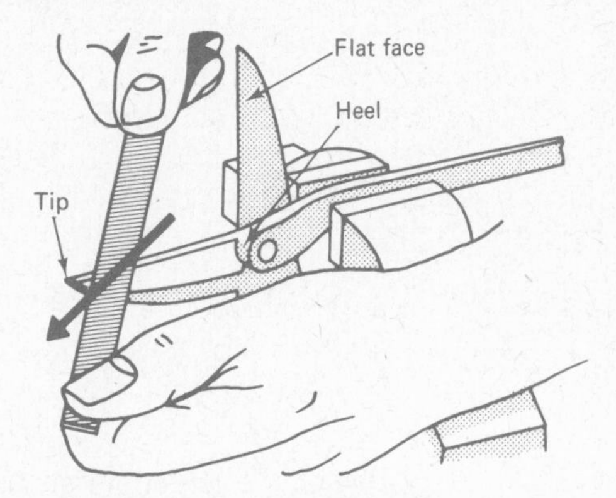

Fig. 1-18. Sharpening tin snips.

1-20. Wire Stripper

Wire strippers are used to remove insulation from solid or stranded wires without placing damaging nicks or breaking strands of the conductor wire. The wire stripper shown in Figure 1-19a is inexpensive and adequate for occasional use. It is 5″ long, has a spring to hold the cutting jaws open, and an adjustable cutting jaw closure for different-sized conductors.

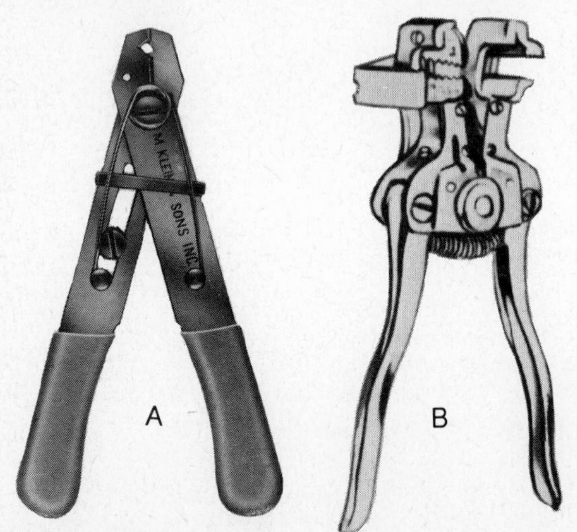

Fig. 1-19. Wire strippers: (A) inexpensive model; (B) model used where extensive wire stripping is performed.

To set the jaw closure to the proper wire gauge opening, loosen the screw and slide the screw, cam and indicator until the indicator points to the gauge of the wire being cut. Then tighten the screw. Place the wire into the open jaws and then close the jaws tightly onto the insulated wire. Pull the wire and strippers

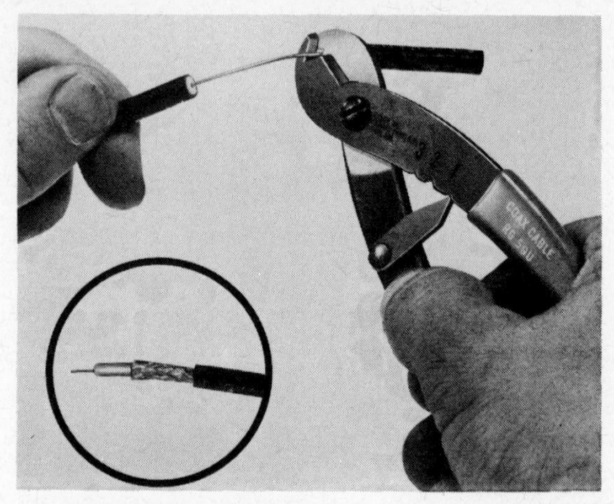

Fig. 1-20. This special wire stripper is used to strip the insulation, shield, and inner conductor insulation from coax cable.

in opposite directions as the insulation is stripped from the wire. Figure 1-20 illustrates a special wire stripper for stripping insulation from coaxial cable (RG-59U). Three notches in the handles provide the correct jaw openings for cutting through and stripping the outside insulation, the shielding and the insulation of the inner conductor.

Figure 1-19b illustrates a wire stripper for use in production work. The wire to be stripped is placed into the proper wire gauge slot, the handles are squeezed and are then released; the wire is stripped. During this operation, the tool grips the wire, cuts through the insulation, and strips the insulation off. An adjustable guide on the head of the tool can be set to remove from 1/4″ to 1″ of insulation for every squeeze of the handle. The hardened jaws of this 7″ stripper are easily replaced. Six openings handle wire sizes from no. 8 through no. 22.

2 Planes

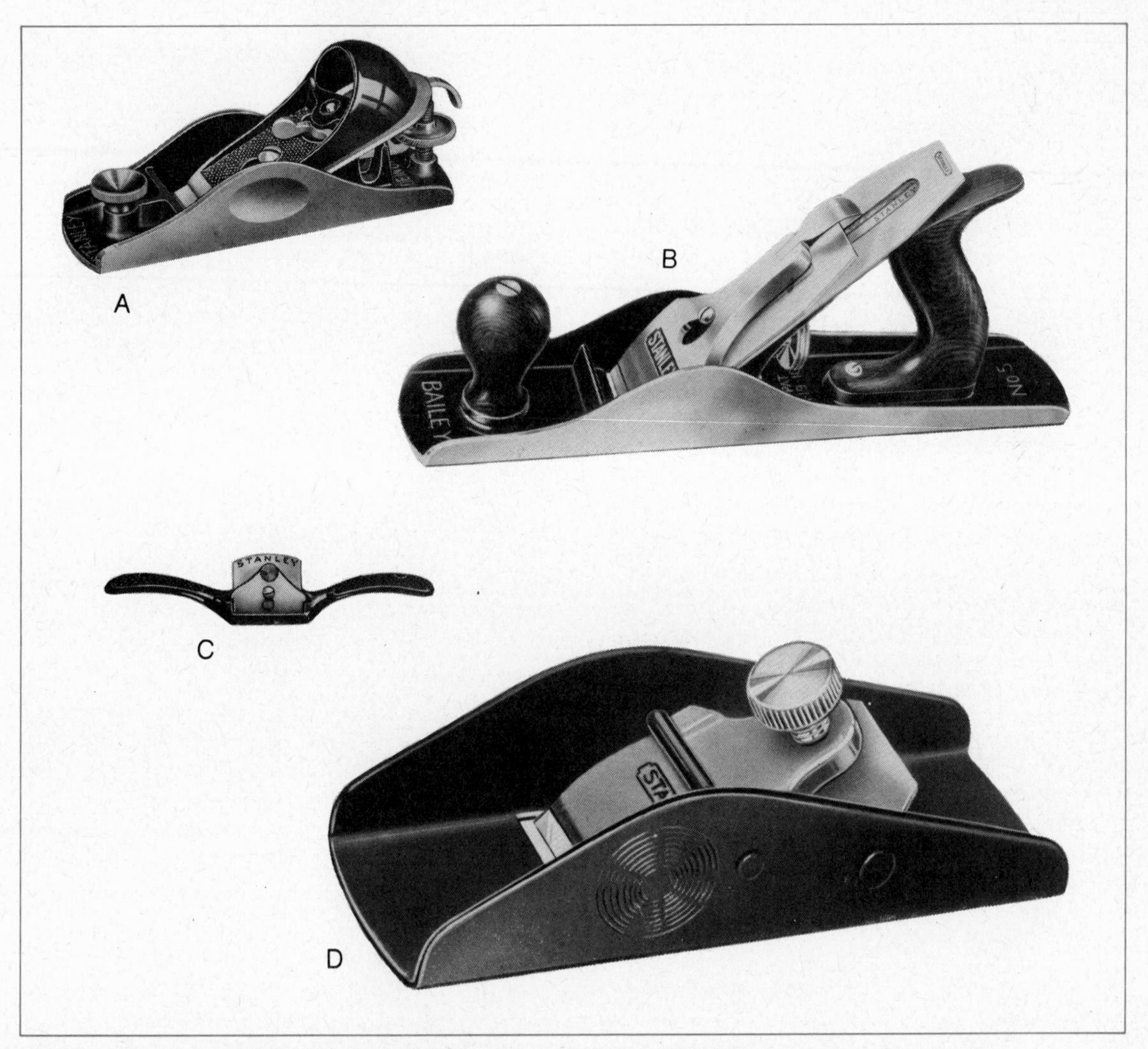

Fig. 2-1. Planes: (A) block; (B) jack (smoothing, fore, jointer); (C) spoke shave; (D) trimming.

2-1. General Description

A plane is a tool used by carpenters, cabinetmakers, home craftsmen and hobbyists to smooth, shape, trim, bevel or chamfer a wood workpiece. The plane is especially useful for removing wood from doors, drawers and other workpieces to ensure proper fit. The plane (Figure 2-2) consists of a chisel-like bevelled blade, cap iron, bottom, handle, knob, adjusting nut and lateral adjusting lever. The tempered steel cutter blade, which has a bevelled, ground and honed cutting edge, is often called the *plane iron.* The cap iron clamps to the blade (plane iron) and is used to curve and break the shavings coming through the mouth of the bottom. The bottom is an iron casting with machined sides and bottom; the bottoms of most planes are flat, but some have grooved bottoms to lessen the drag when they are used with green and gummy wood. Handles and knobs of the larger planes are made of hardwood.

The planes discussed in this chapter are those that are used most often: the *block* plane, employed primarily for planing the end grain of a piece of wood; the *jack* plane (smoothing, fore and jointer), used to plane surfaces and edges of wood to a smooth finish; the *trimming* plane, applied usually by hobbyists to shape small workpieces; and the *spoke shave,* used to smooth curved surfaces.

Assume that you have a long workpiece with a wavy edge that you'd like to shape to a smooth, flat surface. Which plane should you select? You should select the plane having the longest body because a small plane (such as a block plane) tends to ride up and down irregular surfaces. Long planes like the jack, fore and jointer planes trim the peaks off of irregular edges. Thus, the longer the plane, the better it is for trimming peaks. Since the jack plane is probably the most adaptable to different types of jobs, we recommend it as your first plane. The block plane is the second most needed plane.

One other tool that could be mentioned in the plane category is the *forming tool.* The forming tool consists of a holder and a blade that is made up of hundreds of curved teeth that are actually miniature milling cutters (see Section 6-6, Volume 3).

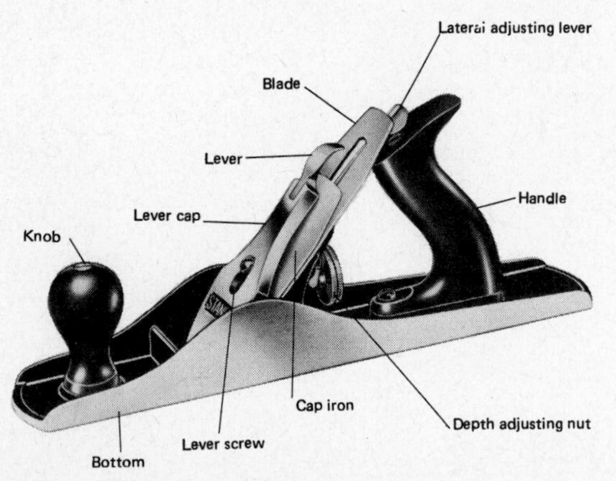

Fig. 2-2. Parts of a plane.

2-2. Application of Planes

The first step in the correct use of a plane is to set the blade: it must be straightened laterally and set at the desired depth. Hold the plane upside down with one hand and sight from the front into the blade. On jack, smoothing, fore, jointer and adjustable block planes, move the lateral adjusting lever until the blade is straight. On block planes that do not have a lateral adjustment lever, loosen the adjusting wheel and set the blade straight by hand; also set the blade depth and retighten the wheel. If the blade of a plane is not straight, thicker wood cuts will be taken from one side of the workpiece than from the other.

Rotate the depth-adjusting nut to move the blade to the correct depth. Set the blade deep for rough cuts and shallow for fine cuts. The plane can cut shavings as thin as the pages of this book.

Make sure the cap iron screw is tight so the cap iron is right against the blade. Also be sure that no wood chips are wedged between the cap iron and the blade. The cap iron should be only 1/16″ from the end of the blade for planing softwoods – less than 1/16″ for hardwoods. The cap iron breaks the curled shavings coming up through the mouth of the plane.

With the plane properly set for cutting, secure a workpiece into a vise, clamp or any other improvised holding device (see Sections 4-6 and 4-17, Volume 3). The best height for the workpiece is slightly above your waist. For practice, use a piece of scrap wood of the same species as your workpiece. Grasp the plane handle with one hand and the knob with the other. Place the front of the plane (the area in front of the blade) on the workpiece. Level the plane and hold it square (parallel to the floor) to the workpiece.

Move the plane with the grain of the wood. Starting at one end of the workpiece with the blade off the end, push the plane across the board with a long, continuous, even stroke. Keep the pressure constant. At the beginning of the cut, apply a little more pressure to the knob end of the plane than to the handle. When the plane is on the board, apply equal pressure to the handle and knob. At the completion of the cut, when the front of the plane moves off the board, ease up on the pressure on the knob. By regulating pressure on the knob at the start and at the conclusion of each stroke, you will keep the ends of the workpiece from becoming rounded.

One effective method of learning to plane correctly is to draw several parallel lines on the edge (along the grain) of a piece of scrap wood. Set the plane to remove very thin strips of wood. Plane along the edge and check to be sure you are planing straight and removing the same amount of material from both sides. You can also draw parallel lines on the top and bottom surfaces of a workpiece near one edge. Plane down to the parallel lines.

You will sometimes get better cutting action if you move the plane at a lateral angle of about 15°. Check the squareness of the cut in all directions often (see Section 9-8, Volume 3).

To chamfer or bevel the edge of a workpiece, first mark the wood for location and angle of the cut. Tip the plane to the desired angle and, holding the angle steady, plane with continuous, straight strokes along the grain of the workpiece.

End grain cutting (board edges) must be done carefully to prevent the end of the board and surface fibers of the workpiece from splitting. There are several practical methods to eliminate splitting: (1) bevel the edges to 45°; (2) sandwich the workpiece between two pieces of scrap wood and plane all three pieces at once; (3) angle the plane laterally at 15° as it is pushed across the end grain; (4) plane from the ends toward the center and then plane the hump out of the center. End grain planing is usually done with a block plane.

2-3. Care of Planes

The plane is only as good as the cutting edge of its blade – the sharpness of the blade must be preserved by frequent honing and occasional grinding when the blade is nicked or completely dulled.

Honing is done on an oilstone that is heavily coated with light machine oil. The blade is held at a 30° angle. The simplest method of holding the blade steady at 30° is to purchase a honing jig that clamps the blade into place. The jig is rolled along the oilstone as the edge is honed. You can also improvise a jig by using a bolt, a nut and a wing nut. Choose a bolt that is long enough to obtain a 30° angle. Clamp the blade between the nut and the wing nut. After honing the edge, rub the back of the edge very lightly in a figure eight direction to remove the burr left by honing.

To regrind a planing blade, adjust the grinder tool rest (or improvise a guide) at an angle of 25°. Hold the blade firmly at the 25° angle and move the blade back and forth all the way across the grinding wheel; this method prevents wear on the wheel. Keep the blade cool by dipping it frequently in water.

When reassembling the cap iron to the blade, place the cap iron slightly back from the cutting edge of the unbevelled side of the blade – about 1/16″ is correct for planing softwoods, a little less than 1/16″ for hardwoods.

Plane blades, as well as all other plane parts, are replaceable on tools manufactured by the large, well-known, reputable tool manufacturers.

Keep the plane blades thinly oiled with light machine oil. When a plane is not in use, lay it on its side to protect the blade. When storing the plane, retract the blade. Keep planes from coming into contact with other tools.

2-4. Block Plane

The smallest plane for general shop use is the block plane. It may be used to cut edges of small workpieces, but its main use is in cutting across the end grains of wood. The plane is constructed so that the cutter angle is small (between 12° and 21°), which enables the tool to cut end grains more easily. It can be used with one hand, but for better control, use two.

There are several styles of block planes available. The one shown in Figure 2-1a has a depth adjusting nut and a lateral adjusting lever. In essence, the block plane illustrated is a small smoothing or jack plane (see description below). Another type of block plane, less expensive than the one shown, has a wheel that turns a screw and tightens the blade within the plane frame. When the wheel is rotated to loosen the blade, the blade depth and lateral angle may then be set by hand. The wheel is then turned to clamp the blade into the desired position. Block planes are from 6″ to 7″ long with cutter widths from 1-3/8″ to 1-5/8″. The blade is mounted with the bevel of the cutting edge *up* toward the top of the plane.

2-5. Jack (Smoothing, Fore, Jointer) Plane

Figure 2-1b illustrates a jack plane. The smoothing, the fore, and the jointer plane are all similar to the jack plane, the only differences among them being their length and blade width. Their lengths are as follows: smoothing, from 5-1/2″ to 10″; jack, from 10″ to 15″; fore, 18″; jointer, from 22″ to 25″. The jack plane blade is 2″ wide and the large plane blades are 2-3/8″ wide.

These planes are used for straightening surfaces and edges and for trimming. The longer the plane, the better it is for cutting peaks. The jack plane is the plane most often used by the home woodworker.

The blades of the planes in this group are mounted with the blade bevel toward the bottom of the plane. To protect the blade from anything that might come into contact with it, store the plane with the blade withdrawn so that it does not protrude from the bottom.

2-6. Spoke Shave

Spoke shaves, so named because of their original use in shaving wheel spokes, are used to smooth or chamfer convex or concave wood surfaces. Its function is thus exactly opposite to that of a block or jack plane: whereas planes are designed to remove curves and make the workpiece flat, the spoke shave is designed to develop and smooth curves. The spoke shave is grasped with both hands and is pulled or pushed along the workpiece, which should be securely held with a vise or clamp. The spoke shave cuts only in the direction of the wood grain.

The spoke shave is about 10″ long and has a cutter which is 2-1/8″ wide. On one model, the blade depth and angle are set by hand after a thumbscrew is loosened; on another model, two screws are provided to set the blade depth and angle.

2-7. Trimming Plane

The trimming plane is a small, lightweight plane, usually from 3″ to 4″ long with a 1″ cutter. It is used by hobbyists and craftsmen to plane small pieces of wood used in models or delicate workpieces. A single knurled nut is used to clamp the blade to the desired depth of cut.

Before starting a cut with a circular saw, pull the trigger and let it reach full speed.

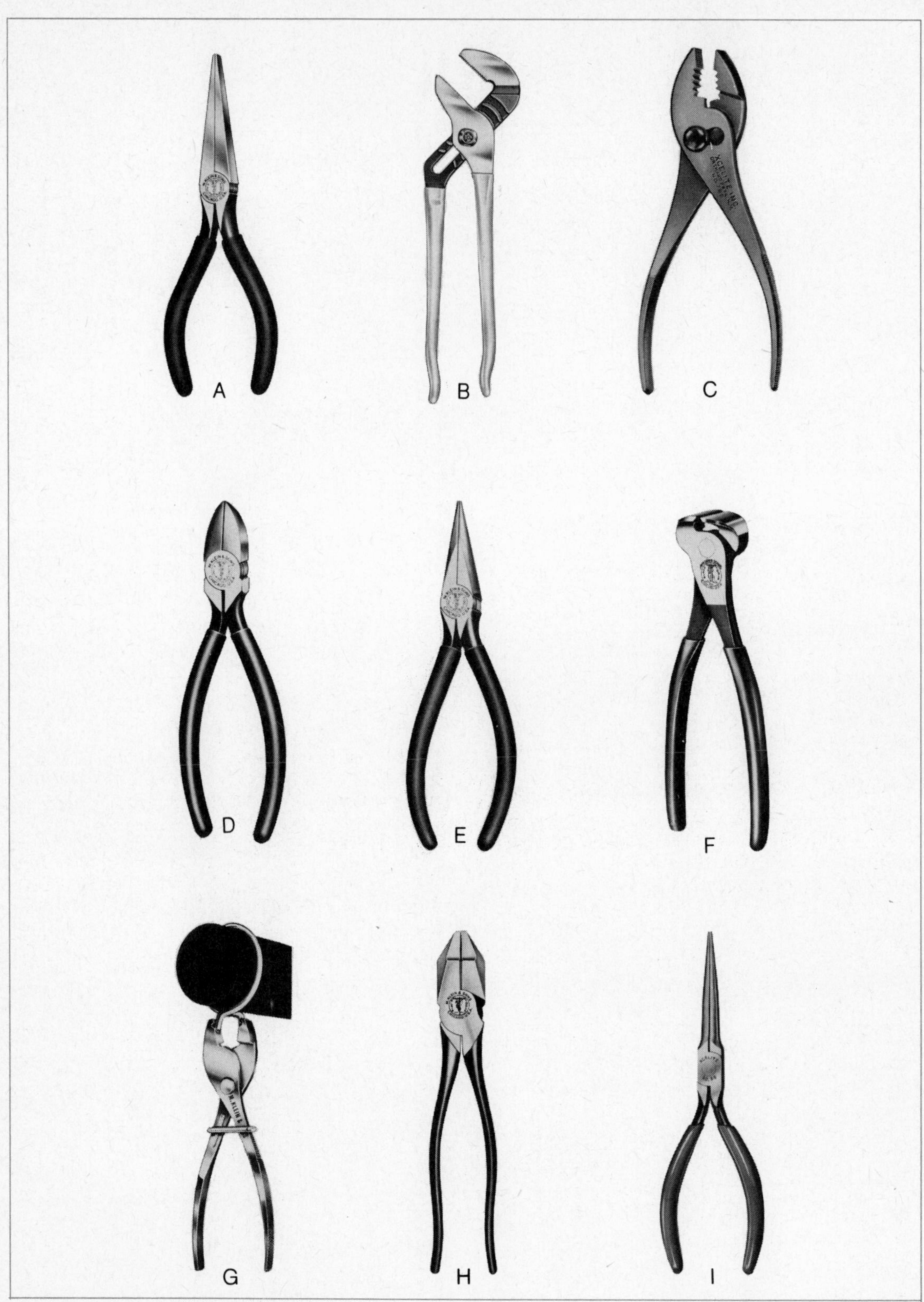

Fig. 3-1. Pliers: (A) chain-nose; (B) channel; (C) combination slip-joint; (D) diagonal; (E) duck bill; (F) end-cutting; (G) hose clamp; (H) lineman's; (I) needle-nose.

Pliers

3-1. General Description

Pliers are the most commonly used tools for gripping, cutting, bending, forming or holding metals or other materials. They can also be used to strip insulation from wire. There are numerous types, sizes and shapes of pliers available, many with a very specific function. The most frequently used types of pliers are *chain-nose, channel, combination slip-joint, diagonal, duck bill, end-cutting, hose clamps, lineman's* and *needle-nose.*

Pliers consist of a pair of milled jaws, a pivot point and a pair of handles. They are made of heat-treated tool steel and are often chrome-plated to prevent rust. Jaws are made narrow to fit into tight places and some have cutting edges for cutting soft wire and stock. The handles are often knurled for a sure handgrip and a coil spring may be placed between the handles to hold them open. The size of the pliers is determined by its overall length, usually between 4″ and 10″.

Midget pliers are only about 4″ long and are used for close work, such as holding, bending and cutting fine wires in electronic, radio and TV work, and in making and repairing jewelry. Pliers of 5″ to 6″ are for general all-purpose work. Pliers of 7″ or larger are for heavy-duty work.

For most shops, one pair each of channel, combination slip-joint, diagonal and needle-nose pliers is sufficient. A locking plier wrench is also useful (Section 8-8).

3-2. Application of Pliers

To hold a workpiece with pliers, place the object as close to the pivot point as possible; this allows the gripping hand to squeeze the handles with the greatest force. If the pliers have a coil spring between the handles, hold them as shown in Figure 3-2; if there is no spring, hold the pliers as shown and place the little finger on the inside and against the plier handle. By opening and closing your fingers, you can control the pressure at the jaws. When using pliers for electrical work, be sure to cut the electrical power. As a precaution, insulate the plier handles with special metal handles; the handles may also be wrapped with several layers of electrical insulating tape.

To cut a piece of wire or stock with pliers, place it between the cutting edges of the jaws so that the wire or stock is perpendicular to the edges. Hold the wire or stock with one hand and place the other hand around the ends of the handles. Squeeze the handles firmly.

In bending and forming a metal workpiece, place the workpiece between the jaws as

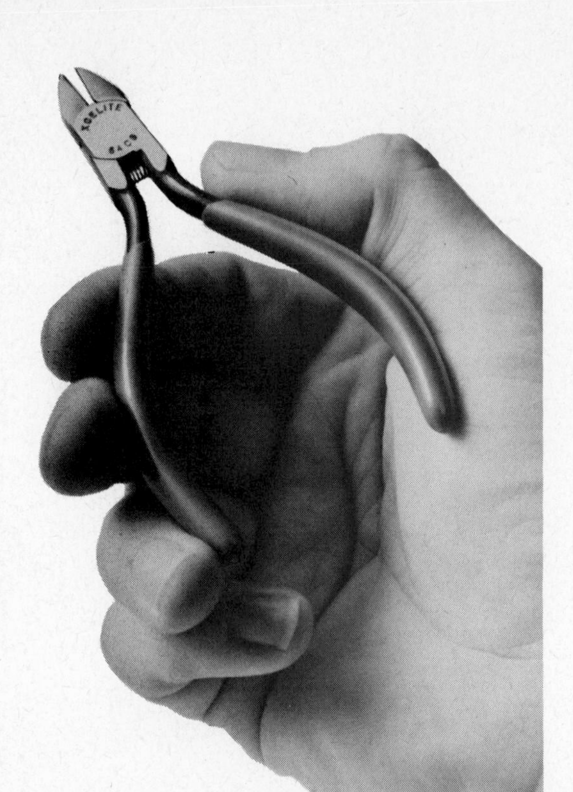

Fig. 3-2. If the pliers have a spring between the handles, grasp them as shown. If there is no spring, place the little finger in the inside of the handle against the handle.

close to the pivot as possible; be sure that no part of the workpiece comes into contact with the cutting jaws. Hold the handles tightly closed and bend and form the workpiece into the desired shape.

When stripping insulation from wires with pliers, place the wire between the cutting edges or notches, with about 1/4″ of wire extending through the pliers. If more than one notch is available, choose the proper notch for the gauge of the wire being stripped. Close the handles lightly so that the edges cut into the insulation; rotate the wire in the cutting edges to *score* the insulation. Pull the wire slowly, stripping the insulation off. Inspect the wire to be sure that the plier cutting edges did not damage it. (Special tools are available for stripping insulation – see Section 1-20.)

You can also use pliers as clamps or as dissipators of heat from a joint when soldering. To use them as clamps, place a heavy rubber band or several small rubber bands around the handles to hold the jaws closed around the workpiece. If you want to protect a workpiece from damage from the serrated jaws of the pliers, wrap tape around the jaws.

Use pliers only for their intended uses. While you should not try to increase the leverage of plier handles by lengthening the handles with sections of pipe or other extensions, you can slip rubber tubing over the handles to cushion your hands.

3-3. Care of Pliers

Damaged or worn serrated jaws may be repaired by filing. If possible, separate the jaws by removing the pivot nut and screw. Place the plier jaw in the protected jaws of a vise. Restore or repair damaged serrations by filing them with a three-square file.

The cutters on some pliers can be reground. Before grinding, however, inspect the cutter design to determine if the cutting edges will close after material is ground from the edges. Do not attempt to sharpen pliers (such as diagonal pliers) that are not designed for regrinding.

To regrind plier cutting edges, separate the pliers, if possible, and place them in the protected jaws of a vise. Grind the cutting edges so that the reground bevel is parallel to the old bevel. Grind the same amount of stock from both jaws. Cool the jaws frequently in water to prevent loss of temper. (Loss of temper makes the cutters practically useless.)

Keep your pliers clean by wiping them occasionally with a rag and light oil. Place a drop of oil on the pivot pin and open and close the pliers several times to work the oil into the pin.

3-4. Chain-Nose Pliers

Chain-nose pliers are used for miniature electronic parts assembly, mechanical assembly, wire forming and for holding small

Because of its versatility, the electric drill should be the first power tool purchased for use in the home.

parts. A set of cutting edges to cut soft wire or stock may be located in the jaws near the pivot point. Chain-nose pliers are very similar to needle-nose pliers (Section 3-12) except that the chain-nose pliers are shorter, wider and hence stronger. They are 5-1/4″ to 6″ long, with jaw lengths of about 1-1/8″. Some models have a plated coil spring between the handles to hold the jaws open.

3-5. Channel Pliers

Channel pliers are similar to the familiar combination slip-joint pliers (Section 3-6) except that the jaws stay nearly parallel. This is a decided advantage because it allows the channel pliers to grip large, flat surfaces such as the flats on hexagon nuts used for sink trap plumbing. The channels in one leg of the plier provide for quick, non-slip adjustments. The long handles give additional leverage and reach.

In use, the jaws are positioned parallel at approximately the proper opening and are locked into place by engaging the tongue in the proper groove. The jaws do not slip even under heavy pressure.

Channel pliers similar to those in Figure 3-1, but with slight modifications, are known by numerous names, including power groove, utility, arc joint, mechanic's, pump, joint and long interlocking joint pliers. Lengths of the channel pliers are from 6-1/2″ to 16″.

3-6. Combination Slip-Joint Pliers

The combination slip-joint pliers are used to hold or bend small pieces of flat or round metal stock, tubing, bars, wires and a variety of other items. The slip-joint allows for adjustment of the serrated jaws to a wider opening that gives more closing force leverage to the gripping hand. Some combination slip-joint pliers have a short set of wire-cutting jaws near the pivot pin. The serrations on the end of the jaw are fine; those at the center of the jaws are coarse. The capacity of the jaws is approximately 3/4″ for 5″ pliers, and 1-1/2″ for 10″ pliers.

If the pliers become loose at the pivot, tighten the pivot nut. If the nut continually loosens, retighten it and then, using a punch and hammer, burr the pivot threads so that the nut cannot be loosened.

3-7. Diagonal Pliers

Diagonal pliers are used to cut soft wire, small stock and cotter pins to proper length. Diagonal pliers have short jaws with the cutting edges at a slight angle that enable the diagonals to make close cuts. Some diagonal pliers are notched for wire stripping; the jaws of this tool are not used for gripping. Length is approximately 6″.

3-8. Duck Bill Pliers

Duck bill pliers are named for the similarity of their shape to a duck's beak. They are also similar in shape and function to the needle-nose and chain-nose pliers (Section 3-12 and 3-4). Duck bill pliers are used for mechanical and miniature electronic assembly, wire forming and for holding small parts. Square bends are made with the duck bill pliers, whereas round bends are made with the needle-nose pliers. The lengths of the pliers vary from 5-1/2″ to 6-1/2″, with jaws from 1-3/4″ to 2-1/4″.

3-9. End-Cutting Pliers

End-cutting pliers, known also as *nail-nipping pliers* and as *pincers,* are used to pull nails and to cut wire, brads and nails close to the work. The jaws are 7/8″ long; the overall length of end-cutting pliers is about 7-3/4″.

To cut off wire, brads or nails, place the jaws around the object, grip the pliers at the end of the handles and squeeze until the object is cut. To remove nails, lightly grip the nail with the jaws at the workpiece surface. Hold the pliers at the handle ends so that the rounded side of one of the plier jaws is against the workpiece. Push the handles downward in an arc, prying the nail out of the workpiece. Continue in this manner until the nail is removed.

3-10. Hose Clamp Pliers

If you have changed automobile radiator hoses using combination slip-joint pliers to open the spring tension hose clamps, you'll welcome the hose clamp pliers with notched jaws to keep a hose clamp from slipping. Notches are cut into the jaws for gripping the spring with the pliers in either of two positions, each 90° from the other. A slip ring that rides over notches on the plier handles permits the plier jaws to be clamped in a variety of open positions so that the clamp is open and the hose can be adjusted. Hose clamp pliers are useful in the installation or removal of hoses. The jaw capacity is from 0 to 1-3/4″; plier lengths are about 7-1/2″.

3-11. Lineman's Pliers

Lineman's pliers are used to grip flat materials and to bend, cut and strip insulation from wires. The flat, serrated jaws are used to twist wires together. An insulation crusher is built in. Different-sized lineman's pliers are often referred to as *electrician's* and *side-cutting* pliers; lengths vary from 4″ to 9″.

3-12. Needle-Nose Pliers

Needle-nose pliers (also called *long-nose pliers*) are used for mechanical assembly, miniature electronic parts assembly and wire forming; they also hold washers, nuts, or other small parts, and they shape jewelry. They are especially useful for reaching into tight spaces. Some models have jaws curved around to an angle of 90° from the handles. The needle-nose plier jaws may also have a set of cutting edges near the pivot point for cutting soft wire or stock. Some models have a plated coil spring between the handles to hold the jaws open. Plier lengths are 6″ to 8″, with jaw lengths from 1-1/2″ to 2-1/2″.

When buying needle-nose pliers, make sure the jaws align properly when closed.

The radial arm saw may be used for making dadoes and for quick, accurate crosscuts.

The table saw is used for cutting large pieces of stock.

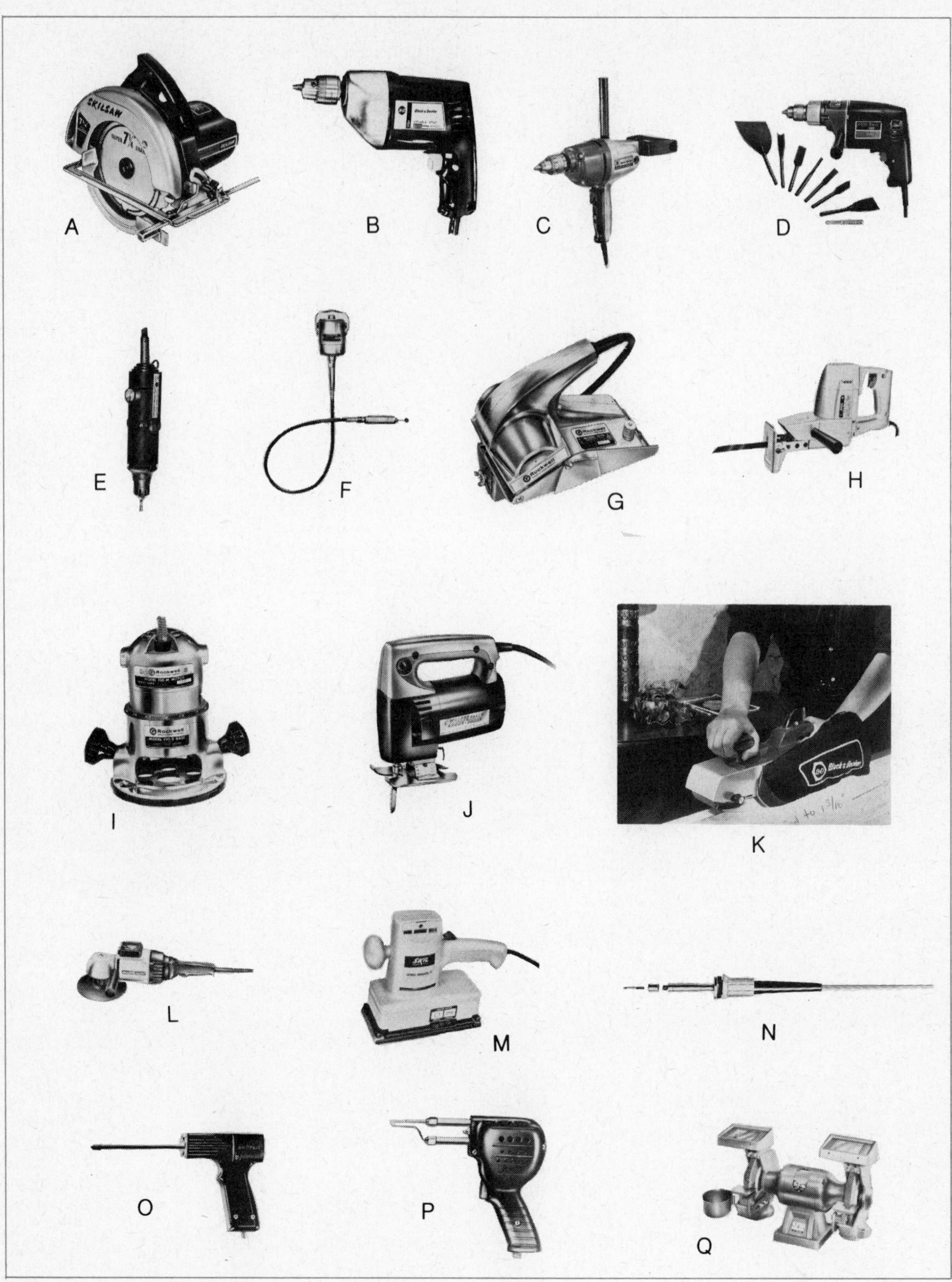

Fig. 4-1. Powered hand tools: (A) circular saw; (B) electric drill; (C) heavy-duty electric drill; (D) multipurpose drill; (E) modeller's power tool; (F) modeller's power tool with flexible shaft; (G) plane; (H) reciprocating saw; (I) router; (J) sabre saw; (K) belt sander; (L) disc sander; (M) finishing sander; (N) miniature soldering iron; (O) pistol-grip soldering iron; (P) instant heat soldering iron; (Q) bench grinder.

Hand-held Power Tools

4-1. General

Power hand tools perform many of the same functions as the hand tools described elsewhere in these volumes, buy they do so more efficiently and often more accurately. Power tools also increase your range of capabilities and save your time and energy. And once you master the fundamental operation of the basic tool, you can extend the tool's usefulness by adding accessories.

As with any tool, a power tool's principles of operation, its use, capabilities, limitations, safe operating procedures, and general care are learned only through practice. We suggest that before you use a tool for the first time, you spend a few hours practising with it. First, read the complete section of this chapter dealing with the tool. Then read the manufacturer's instructions. Pick up the tool; learn the names of its parts. Hold the tool in position. Move its switches. Move its adjustments. Note its safety features and any conditions where its use may be unsafe. Locate lubrication points. Now it is time to move to the shop.

Clean up your shop. Make room for your new tool. Get some scrap wood and secure it to your bench. Make sure that all adjustments on the tool are secure. Plug your tool into the power outlet and turn it on without touching the workpiece. Respect the power in your hands. Turn the power off. Place the tool to the scrap wood, apply power, and practise using it to perform its basic functions. Try more difficult operations. Extend your knowledge of the tool's capabilities and limitations by varying adjustments, changing angles, or using different cutters. Add accessories to the basic tool and practise with them. Practise. Practise. Practise - until you are thoroughly familiar with the complete operation of the tool. Now you are ready to use the power tool on a workpiece.

The power hand tools covered in this chapter include the *circular saw, electric drill, grinder, modeller's power tool (*and *accessories), planer, reciprocating saw, router, sabre saw, sanders (belt, disc, finishing), soldering irons* and *x-tra tool.* Read over the sections describing those tools you have. Skim over the descriptions of those tools that you don't have to gain some knowledge applicable to the tools that you do have. The new knowledge will guide you in buying the next tool you want.

When purchasing power hand tools, wait for a sale so that you can save money, but *do buy high-quality tools* from reputable manufacturers. Be careful if buying multi-tools or tools with many accessories because merchandisers may take the opportunity to sell you items that no one else wants and that you'll probably never use.

If possible, buy tools from a manufacturer that has a repair facility or representative

The jigsaw is especially useful in cutting square, circular, irregular or intricate cutouts.

located in your city. Before you buy, be sure that you can obtain replacement parts for the tool. Also be sure the tool has a guarantee – and don't forget to mail it in. Finally, don't buy a tool unless a set of operating and maintenance instructions and a list of repair parts are included. These suggestions are particularly important if you are buying a used tool.

When purchasing power tools, be sure that you buy a size with a horsepower rating and speed sufficient to accomplish your jobs. Don't consider only the job at hand, but think also of future tasks. Tools with ball bearings outlast other tools. Motors with slip-clutch gear trains allow the motor to operate even if the drive shaft becomes suddenly jammed. This prevents gear stripping and motor burnout. (If a drive shaft stops rotating because of a jammed blade, immediately cut off the electrical power.)

For safety, you should buy only power tools having one of two wiring features: a three-wire grounded system or a double-insulated tool. These features protect you from electrical shock in case of internal electrical damage. The three-wire grounded system is used with three-prong plugs. If your shop doesn't have a three-wire system, rewire it now – for safety's sake. When using three-wire power tools in other parts of the home, use a properly installed adapter plug. On double-insulated tools (two-wire power), the outer shell of the tool is completely insulated from the wiring.

When you are selecting cutting blades, keep in mind that tungsten carbide blades cost more than standard blades, but they outlast the standard blades ten to one. Instead of teeth, tungsten carbide blades have particles of tungsten carbide fused to the blade. These blades cut wood, ceramic, counter top material, slate, cement, brick, asbestos, pipe and stainless steel.

4-2. Application of Power Tools

The application of each tool is described separately, but here are some general tips applicable to all power tools:

1. Always be safe. Be sure the workpiece is firmly supported and clamped. Assume a safe operating position: make sure you can reach the entire workpiece without losing your balance. Place power cords out of the way; add an extension cord if necessary. Clean the shop of all clutter and leave yourself ample space for working. Be aware of what to do should an accident occur.

2. Try operations on a piece of scrap before beginning on the workpiece. That way you can check out your procedures, adjust the tool (the angle, depth of cut, speed and feed rate) and possibly prevent yourself from making an error and ruining your workpiece.

3. When cutting, use slow speeds for hard materials, and high speeds for soft materials.

4-3. Care of Power Tools

The special care of each power tool is discussed in the appropriate section, but here are a few general tips applicable to all power tools:

1. Brush dust away from the tool after each use. Occasionally use a vacuum cleaner to clean dust from around the motor (hold the vacuum nozzle close to the cooling fins of the motor and draw the dust out).

2. Remove gum, pitch and dirt from tools and accessories with a rag dampened with turpentine.

3. Check power cords for frays, cracks or cuts. Replace the cords in accordance with the manufacturer's instructions if any defects are found.

4. Apply oil (or other lubricant) to the tool motor and other moving parts as recommended by the tool manufacturer. Remove and replace old grease as recommended.

5. Inspect the motor brushes regularly for wear. Excessive sparking indicates worn brushes or springs. Remove and replace the brushes and springs in accordance with the tool manufacturer's recommendations. When new brushes are installed, contour the ends of the brushes to fit against the commutator. Do this by placing a fine abrasive paper against the commutator and lightly rubbing the brush against the abrasive paper. Check the commutator too. If it is dirty, rub it with a fine abrasive paper and then wipe with a rag slightly dampened with lacquer thinner. Finally, brush and vacuum out all dirt.

4-4. Circular Saw

The circular saw is used to crosscut or rip wood, masonite, fiberboard or other similar materials into two or more parts. In addition, it is used to cut mitered joints, to dado and to bevel edges. It cuts along straight lines only. The circular saw is used primarily to make rather rapid cuts that may be slightly rough, but if used carefully and properly, it can make smooth, accurate cuts. It is almost a necessity to use a circular saw if you frame a cottage, add a new room or remodel an existing room. Its light weight and small size make it portable.

A circular saw is labeled by the size of the blade that it uses. Thus a 7-1/4″ circular saw uses a blade having a diameter of 7-1/4″. Circular saws are available in 6-1/2″, 7-1/4″ and 8″ sizes. They are also characterized by the horsepower (HP) and revolutions per minute (RPM) of the motors, which typically are from 1 to 2-1/2 HP and 4700 to 5200 RPM.

A 7-1/4″ 1-1/2 HP 5200 RPM saw is more than adequate for most home building uses. The important consideration in determining the size you need is the depths (thicknesses of wood) to which a saw will cut at 90° and 45° angles. These are listed in Table 4-1.

Nearly all circular saws have the same features incorporated in them, so selection depends on your personal preference regarding manufacturer and style. Circular saws feature a trigger grip on-off switch, sawdust ejection through (or around) a retractable blade guard, an adjustable rip fence with an indicating scale, a cutoff guide, a vertical adjustment to control the depth of cut, and a tilting shoe (base) with a graduated scale for making bevel cuts.

Before plugging in the circular saw cord, install the correct blade for the material to be cut (blades are discussed below), set the depth adjustment, the bevel adjustment and the rip fence (cutoff) guide, if it is to be used. To install a blade, use a box wrench (usually provided with the saw) to remove the arbor bolt. Place the blade on the arbor so that the teeth at the front of the saw are pointing upward. When you look directly at the mounted blade, it will turn counterclockwise during operation (Figure 4-1a). Reinstall and tighten the arbor bolt. Loosen the depth adjustment; grip the saw shoe and pull it down from the saw. Set the saw so that the depth of the blade and cut will be 1/4″ greater than the thickness of the material to be cut. Tighten the depth adjustment securely. If a bevel cut is to be made, loosen the bevel adjustment; grip the shoe and tilt it until the angle required is indicated on the bevel scale. For accurate work, set the angle with a protractor. Tighten the bevel adjustment. If you are making an end cutoff or are ripping a board, loosen the rip guide adjuster, set the

Table 4-1. Circular saw depth cuts

Saw size (in.)	Depth of cut at 90° (in.)	Depth of cut at 45° (in.)
6 1/2	2 1/16	1 3/4
7 1/4	2 3/8	1 7/8
8	2 13/16	2 1/4

guide to the proper dimension and tighten the adjuster.

Now carefully check that all adjustments are proper and tight. Plug the power cord into an electrical socket. Grip the circular saw knob with the left hand and the handle with the right hand (sorry, there aren't any left-handed circular saws for lefties). The index finger grips the trigger on-off switch. Momentarily squeeze and release the trigger. Be sure the saw blade is travelling in the right direction. If all is well, you're almost ready to cut. But first, study the following sawing procedures:

1. Firmly support the workpiece on saw-horses (Section 4-14, Volume 3). Position the workpiece with the *best surface facing the floor* in order to ensure that any wood splintering will occur on the rough or backside rather than on the best surface. Position the scrap to be cut off to your right.

2. If necessary, mark the cutting line – no line is required if you are using the rip fence guide.

3. Stand out of the line of the cut – keep to the left of the cut.

Fig. 4-2. Use the rip fence guide as a cutting aid in sawing straight lines.

Fig. 4-3. Keep the circular saw shoe flat against the workpiece.

4. Place the front of the saw shoe flat onto the workpiece with the saw on the part you are cutting. Don't place the saw on the scrap wood.

5. Slide the saw so that the rip fence guide lies along the edge of the workpiece (Figure 4-2), parallel to the cut to be made. Note that this guide is only usable for cuts about 6″ to 8″ wide. If the guide isn't to be used, place the blade just to the right (in the scrap area) of the cutting line.

6. Pull the trigger and let the saw reach full speed before starting the cut.

7. Holding the saw with the shoe level on the workpiece (Figure 4-3), push the saw along the saw line (or hold the rip fence guide along the edge). Note that as the saw starts into the cut, the blade guard retracts.

8. Keep the blade on a straight line. Don't twist the saw and bind the blade, since this can cause it to kick back. Hold the saw firmly. Cut as fast as the blade will cut.

The router may be used to make rabbet and dovetail joints, dadoes, grooves and bead moldings.

Don't force the saw so much that its speed is appreciably slower.

9. The saw line cutting guide is no longer useful after it runs off the end of the workpiece. Watch the blade for the last inch or so of cutting. Hold the saw steady as the scrap falls off.

The following hints provide you with additional information for making accurate saw cuts:

1. To make accurate, straight cuts, clamp a piece of straight-edged scrap wood to the workpiece. Run the saw (left side of the saw) along the edge. This technique may be used for making crosscut, rip, miter or combination (mitered and bevelled angle) cuts.

2. Tilt the shoe to make bevel cuts (Figure 4-4). For accurate bevel angles, check the shoe with a protractor.

3. To make a shallow cut (not completely through the board), raise the blade and secure it into position. Check the depth of cut on a piece of scrap wood before cutting the workpiece.

Fig. 4-4. Adjust the circular saw bevel adjustment for bevel cuts.

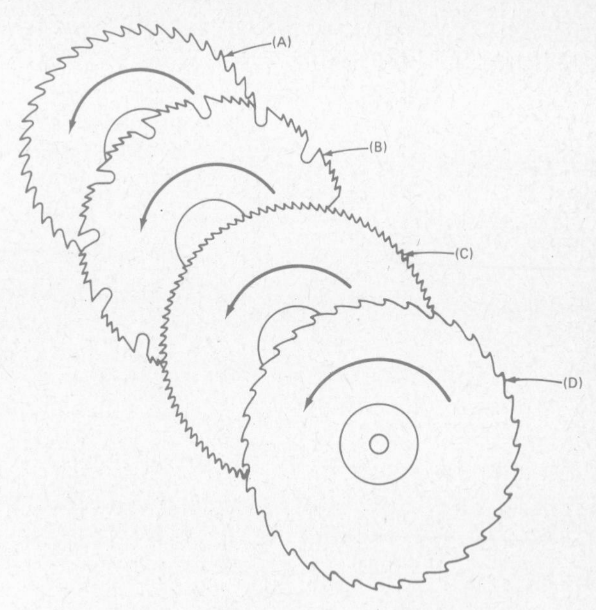

Fig. 4-5. Circular saw blades: (A) all-purpose; (B) planer; (C) cut-off; (D) rip.

4. To dado, make several shallow cuts. You can use a number of cuts to clean out the groove or you can use a chisel to remove the wood between several shallow cuts.

5. If you want to cut a notch in a workpiece, remember the saw kerf. One surface of the workpiece is cut more than the other surface because of the round blade. Finish the notch with a handsaw.

6. Plunge cuts can be made into the center of a workpiece. (A plunge cut is a cut made through the surface of a workpiece rather than at the edge.) Secure the workpiece. Start with the saw blade out of the wood. Firmly place the front of the saw shoe against the workpiece. Hold the back of the saw in the air. Manually retract and hold the blade guide. Start the saw. When it is up to speed, lower the saw slowly and carefully into the workpiece. When the blade has passed through the workpiece, release the blade guard and let it snap into position. Continue your cut.

7. Use sawhorses to support large workpieces.

It goes without saying that a properly selected, sharp blade is required for safe, accurate, efficient cutting. *All-purpose* blades (Figure 4-5a) are used for all-purpose wood cutting – crosscuts, miters, and ripping. *Planer* blades (b) are used for very smooth cuts in all cutting directions. *Cut-off* blades (c) are used for crosscutting and *rip* blades (d) are used for rip sawing. Special blades are available for cutting plywood, counter topping, metal and masonry.

One attachment that is worth noting is a miter arm attachment. Two types are available: one is portable, the other a stationary model that adapts your hand circular saw to a radial arm-type saw. The circular saw slides accurately along metal tracks. Miter cuts 45° left or right can be made.

Be sure that your blades remain sharp. Dull blades can be resharpened by a professional saw sharpener. (See SAWS – SHARPENING AND REPAIRING in the yellow pages of your telephone directory.)

Occasionally wax the bottom of the shoe to reduce friction as it slides across the workpiece.

4-5. Electric Drill

The electric drill should be the first power tool purchased for use in the home. It was originally named a drill because it did just that – it drove bits that drilled holes. But today, this versatile tool is used as a driving unit and is coupled with accessories to become a disc or drum sander, grinder, buffer, wire brush, saw (reciprocating, hack, and circular), screwdriver, hole cutter, rasp, paint mixer, paint sprayer and nut driver.

Electric drills are described by the maximum capacity of the chuck. Drills are available in 1/4″, 3/8″, 1/2″ and 3/4″ sizes; the first two sizes are used in the home, while the latter two are primarily for industrial or construction uses. Large drills must run at slow speeds because they require more torque than small drills; hence, the greater the drill capacity, the slower the drill operates. Drill speeds are fixed on some models, have two speeds (slow and fast) on others, and are variable on newer models. Motor horsepower increases with size too, because it takes a larger motor to provide the driving torque. Drill size capacity, speeds and horsepowers are summarized in Table 4-2. The maximum variable speed indicated is the approximate speed of a fixed drill.

The standard pistol grip style of drill (Figure 4-1b) is the most popular since its design lends itself to the majority of jobs. On pistol grip drills with capacities of 1/2″ or over, extra hand-holding power is required to keep the drill from twisting during operation. A removable bar located at 90° to the chuck end is attached when the drilling operation requires it. The need for additional forward pushing power and holding control by the operator during heavy drilling operations has necessitated the installation of a permanent handle grip located at the back of the larger drills (Figure 4-1c).

The major parts of a drill are the motor, gear drive, housing, on-off trigger switch and variable-speed control, trigger lock, reversing switch and chuck.

Drills are turned on and off with a trigger switch. On drills featuring variable speeds, the speed is increased as the trigger is squeezed. A trigger lock locks the speed as desired. Variable-speed drills are versatile as they provide high speeds for drilling wood,

Table 4-2. Drill capacities, speeds and horsepower

Drill capacity	1/4 in.	3/8 in.	1/2 in.	3/4 in.
Revolutions per minute	0 to 2250	0 to 1200	0 to 600	250 and 475 – not variable
Horsepower	1/6 to 1/4	1/6 to 1/4	1/4 to 3/4	1 to 1 1/2

Fig. 4-6. Maintain a good handhold on the electric drill during all operations.

sanding, buffing, etc., and slower speeds with more torque for drilling metals, threading holes, and driving screws. Variable-speed control allows you to match the speed to the job.

Many drills have a reversing switch. When placed to the reverse position, the drill spins in the opposite direction. This feature is useful for cleaning out drilled holes, for removing jammed drills and for removing screws (screwdriver bit in chuck).

The drill chuck is a three-jaw attachment that is used to grip an infinite range of round drills and accessory shafts up to the maximum capacity of the drill (actually of the drill chuck). The three-jaw configuration holds the shaft within the centerline of the drill. A tee-

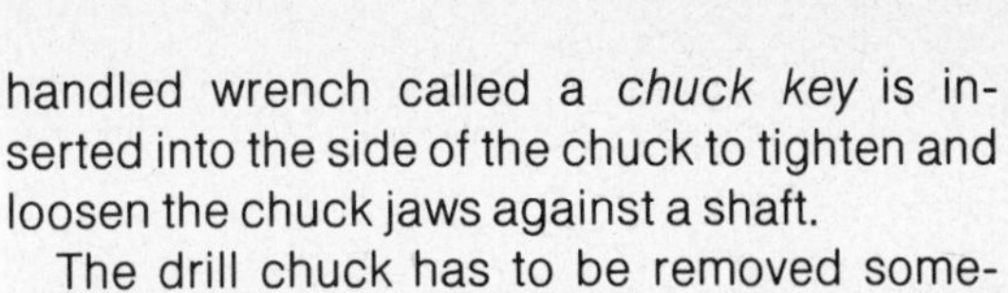

handled wrench called a *chuck key* is inserted into the side of the chuck to tighten and loosen the chuck jaws against a shaft.

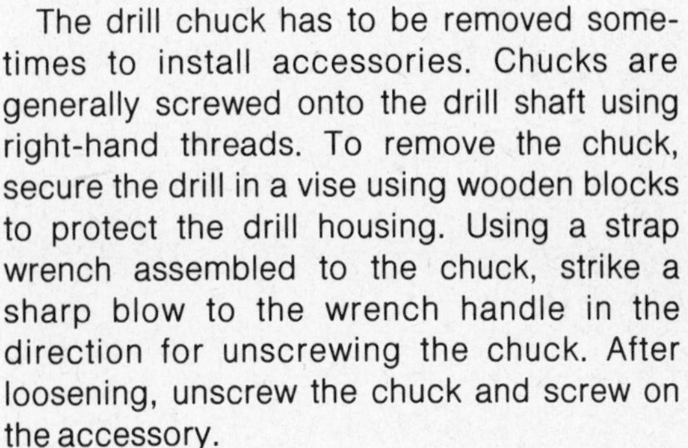

The drill chuck has to be removed sometimes to install accessories. Chucks are generally screwed onto the drill shaft using right-hand threads. To remove the chuck, secure the drill in a vise using wooden blocks to protect the drill housing. Using a strap wrench assembled to the chuck, strike a sharp blow to the wrench handle in the direction for unscrewing the chuck. After loosening, unscrew the chuck and screw on the accessory.

A good all-purpose one-drill workshop should have a 3/8″ variable-speed reversing drill with an infinite speed-locking feature.

When using the drill for any purpose, maintain a good handhold on it (Figure 4-6), as well as a good footing on the floor. Injuries very often occur when a drill jams in a workpiece and the drill twists out of the operator's hands. If possible, secure the workpiece in a vise or with clamps. When using a 1/2″ drill in hard materials, two men should control the drill at all times. (In Volume 3, see Sections 2-2, 2-4, 2-5 and 2-15 for a discussion of drilling techniques.)

Following are some additional tips on drilling. Read them over before you begin your first drilling operation.

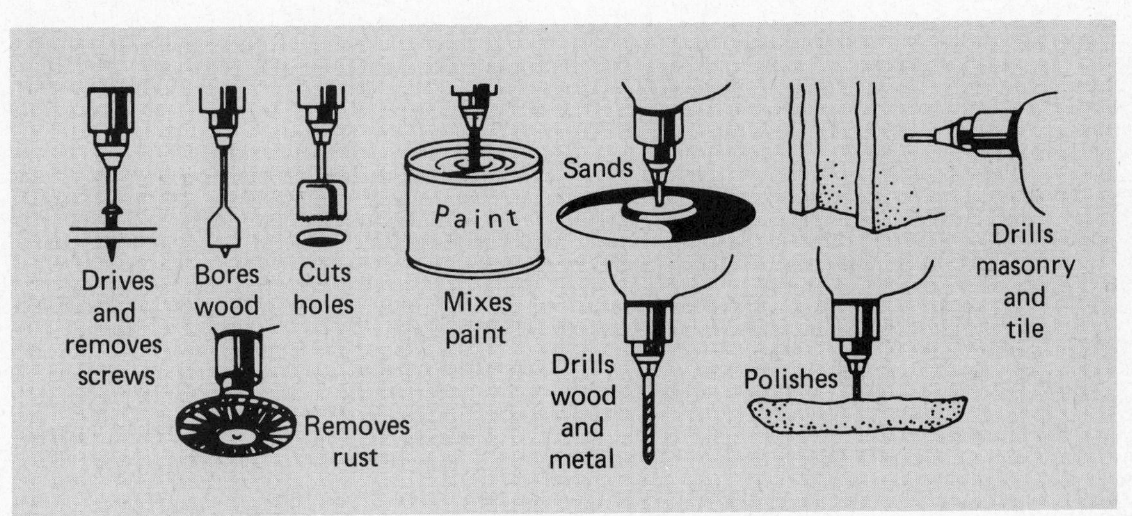

Fig. 4-7. Many accessories are available to extend the use of the electric drill.

1. Before and after inserting a drill or accessory into a drill chuck, inspect the shaft for burrs. Remove any burrs with a file.

2. Insert a drill or accessory shaft into the chuck. As you hand-tighten the chuck, rotate the shaft to ensure that it seats properly in the chuck. Then tighten the chuck and remove the chuck key. Momentarily squeeze and release the trigger and check that the drill or accessory is spinning in the centerline of the drill (i.e. the drill or accessory has been properly seated in the chuck). This procedure is important when you are using small-diameter drills.

3. So that you won't misplace the chuck key, tape the tee-handle to the electric cord of the drill.

4. Use the correct drill speed and feed while drilling or using accessories. Always let the tool do the work. Remember, it might seem slow, but it's a lot faster than doing the work by hand.

5. Appendix K provides decimal equivalents of number and letter sized drills.

The following accessories are available for use on electric drills (the major accessories are illustrated in Figure 4-7): speed reducer, screwdriver attachment, a device to limit the depth of cutting of a drill, drum sander, drum rasp, disc sander, disc rasp, extension drill, polishing wheels, grinding wheels, wire brushes, shaping/router bits, circle cutters, plug (wood) cutters, right-angle drivers for drilling through rafters, jigsaw (reciprocating) attachment, circular saw attachment, paint stirrer, flexible 40″ shaft, carbide-impregnated cutoff discs and a hacksaw attachment. Drill press stands are also available; the electric drill is clamped into the stand, thus becoming a bench drill press.

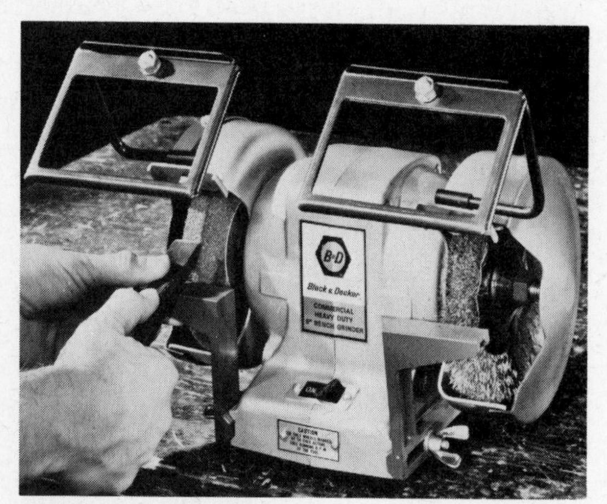

Fig. 4-8. A safe grinder has wheel guards, eye shields, and adjustable tool rests.

4-6. Grinder

The bench grinder cannot be considered a power hand tool. However, it has many applications in the shop with regard to hand tools. It is used to sharpen knives, hatchets, drill bits, chisels, lawn mower blades, garden tools and other shop tools. It grinds off burrs and mushroomed tool heads. With various wheels, it is also used to clean and polish and remove rust.

When buying a bench grinder, be sure it has wheel guard covers and eye shields. These features protect the operator from flying metal pieces and sparks. Grinders should also have tool rests whose distances from the wheel and angles with respect to the wheel are adjustable.

Grinder sizes are specified by the size of the grinder wheels; thus, a 6″ grinder uses a 6″ grinding wheel. Grinders of 5″ to 7″ with 1/4 to 1/2 HP motors are adequate for home use. These grinders run at 3400 to 3500 RPM.

Good grinding wheels have metal bushings that fit tightly onto the grinder spindle. The shaft locknuts should be just tight enough to secure the wheel. After installing a wheel, brush or polishing cloth, stand aside and run the grinder a few minutes to check it out. Turn the power off again.

Adjust the tool rest at the desired angle

1/8″ from the wheel. Be sure that the area behind and around the grinder is clean. Cover your eyes with safety goggles before operating the grinder. Also check the wheel for cracks before applying power; if cracked, replace the wheel. Never use a grinding wheel at a speed faster than the speed marked on the wheel.

When grinding, remove as little material from the workpiece as is required. Keep the workpiece cool by frequently dipping it in water. In renewing a straightedge, pass the workpiece laterally across the grinding wheel (Figure 4-8). The workpiece edge must always be parallel to the grinding wheel.

If the grinder tool rest will not tilt to the desired grinding angle, make a jig to hold the workpiece by hand.

Grinding wheels are made of abrasive grains bonded together. These grains are usually aluminum oxide or silicon carbide. Aluminum oxide is used to grind high tensile strength materials such as high-speed steels and carbon steels. Silicon carbide is used on low tensile strength materials such as brass, bronze, gray iron, aluminum and copper. Wheel grits are categorized as coarse (no. 12 to no. 24), medium (no. 70 to no. 120), and fine (no. 150 to no. 600). Hard wheels are used for grinding soft materials; soft wheels are used for grinding hard materials. A medium wheel is for general use.

Wire brushing wheels are used to remove rust and paint, and to polish. They are available with fine, medium and coarse wire bristles.

Buffing discs are used to polish metals such as aluminum, brass and stainless steel. These discs are made either of lamb's wool or sewn-cotton laminates.

When grinding wheels are in contact with a workpiece, rust, paint, or the material itself sometimes fills the surface of the grinding wheel. This process, which decreases the grinding action, is called *loading* the wheel. When a wheel is loaded, it is cleaned by an operation called *dressing.* Dressing a wheel removes both the load and the dulled grinding grits.

WARNING: When dressing a grinding wheel, always support the star wheel dresser against a bar or plate. Always wear gloves and a full face mask.

The dressing of a wheel is most often performed with a star wheel dresser. This tool consists of a series of hardened star-shaped discs mounted in a holder that allows the discs to rotate freely. The dresser is held with

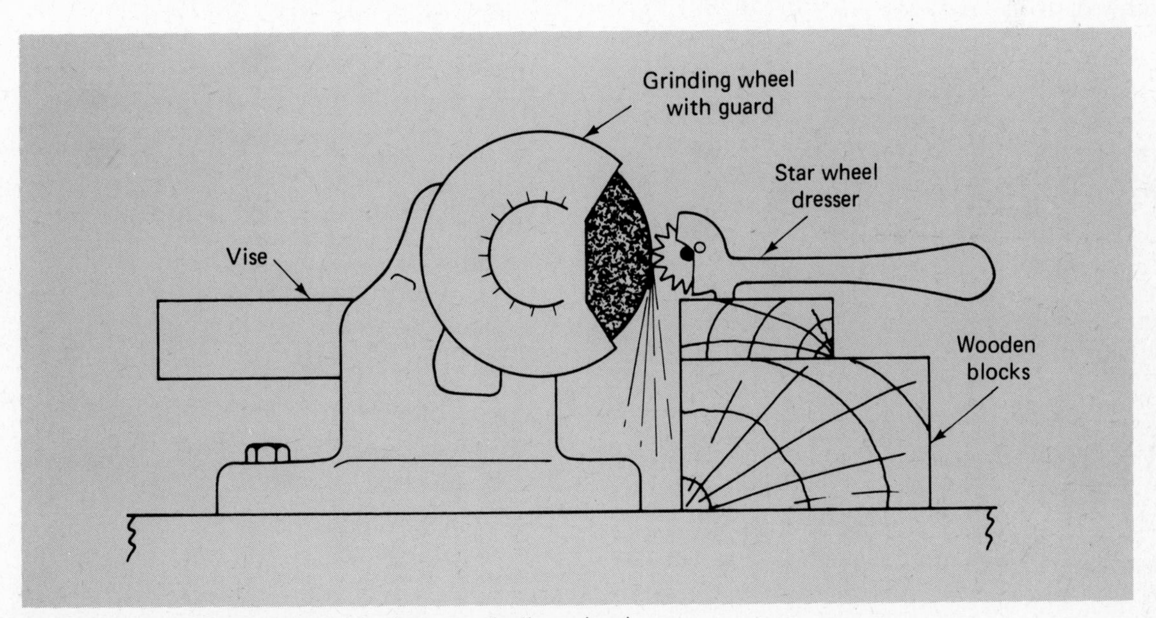

Fig. 4-9. A star dresser is used to dress a grinding wheel.

both hands and must rest against a bar or plate while in contact with the wheel. Failure to support the dresser will cause spin-off from the rotating wheel and possible serious injury to the technician.

The grinder and wheel should be clamped securely in a vise. A support base for the dresser can be constructed of heavy wood near the wheel face to be dressed. The height of the base should put the star dresser approximately on the centerline of the grinder shaft. Make sure that both the grinder and the dresser base are secured. Put gloves on both hands and wear a full face mask.

After starting the grinder, take the star dresser in both hands and rest the head on the support base. Slowly advance the dresser until it makes contact with the wheel. With a straight motion, allow the dresser to traverse the loaded wheel face. This is done until the loaded face is removed and a renewed, flat grinding face is formed. Don't allow grooves to form during dressing. After dressing, always check the flatness of the wheel with a scale.

4-7. Modeller's Power Tool (and Accessories)

Modeller's power tools are virtually miniature power shops. With the numerous accessories available, the modeller's power tools cut metal, saw wood, engrave, drill, polish, grind, deburr, buff, rout and clean. The power tools can be hand-held or mounted in purchased stands or they may be carefully clamped in a vise if protection for the case is provided.

Two types of modeller's power tools are available: the hand-held model (Figure 4-1d) that can be bench-mounted; and the flexible-shaft type (Figure 4-1e). Both styles have valuable features and employ the same tool accessories. Which of the two you choose is strictly a matter of personal preference.

The hand-held modeller's power tool is available with or without a built-in variable-speed control. On models that do not have the speed control, an on-off switch is located on the end opposite the cutter. Always mount or hold the tool so that this switch can be reached readily in any emergency. On variable-speed models, a dial is provided to select the power off position or a variable speed between 5000 and 25,000 RPM. A variable-speed control is essential if the miniature power tool is to reach its maximum efficiency. If this feature is not available on your model, you can purchase an external variable-speed control that mounts on your bench (and you can use it not only for this tool, but for others as well).

Accessories are held in the hand-held modeller's power tool by means of a *chuck*

Fig. 4-10. The drill press stand allows you to drill accurate holes to specified depths.

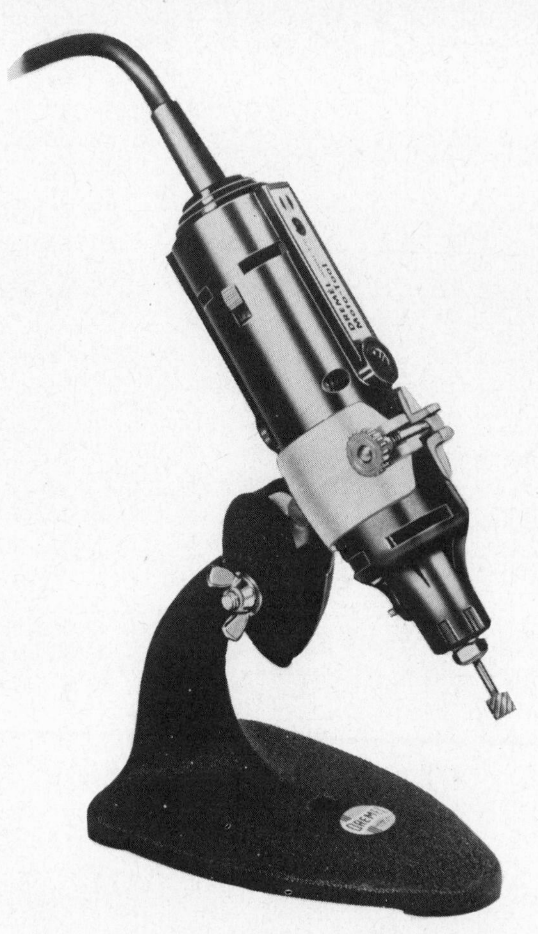

Fig. 4-11. The universal stand holds the tool at any angle.

cap and *collet* (various sizes of collets are used depending on the diameter of the accessory shafts). To change tools, remove the power cord from the electrical socket, and depress and hold the *chuck lock pin* located at the shaft end of the tool. Manually rotate the chuck cap until the lock pin snaps into place and stops the chuck cap from turning. While continuing to depress the lock pin, loosen the chuck cap with the chuck wrench supplied with the tool. When the chuck cap is loose, pull the accessory shaft from the collet. If a smaller or larger accessory shaft is to be used, remove the chuck cap and replace the collet with another of the proper size. Replace the chuck cap. Insert the accessory shaft as far as it will go into the collet. With the chuck lock pin still depressed, use the wrench to tighten the chuck cap. Once the cap is tight, remove your finger from the chuck lock pin and ensure that the spring-loaded pin has come out of the locked position. *Do not start electrical power while the chuck lock pin is depressed.*

The flexible-shaft modeller's power tool is available in a hang-up model or in a bench model. Foot-operated or hand-operated variable-speed controls are available for regulating speeds during operations. Various interchangeable handpieces provide for speeds up to 35,000 RPM.

Accessory shafts are held in the flexible-shaft modeller's power tool by means of a three-jaw chuck. A tee-handled chuck key is inserted into the side of the flexible-shaft chuck to tighten and loosen the chuck jaws against a shaft.

The hand-held modeller's power tool can be mounted in one of two fixtures. The drill

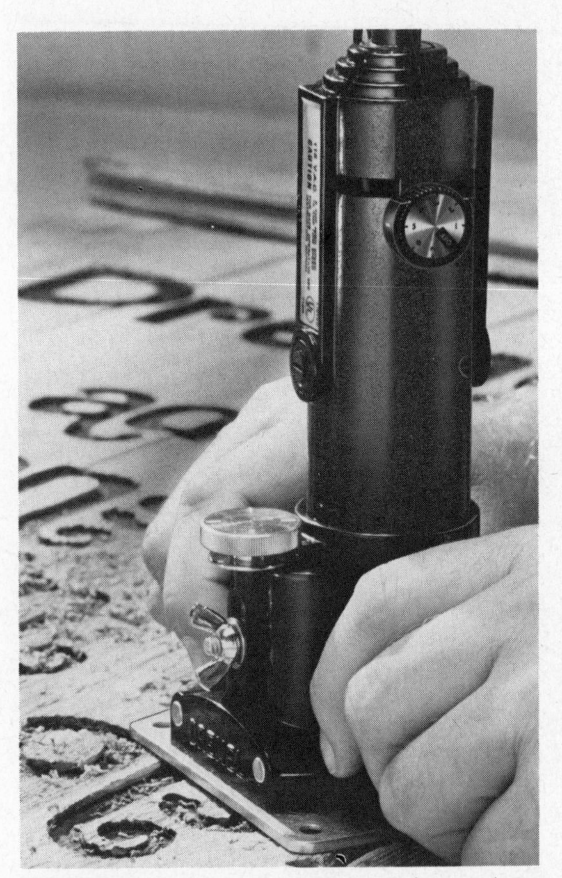

Fig. 4-12. The router attachment can be used freehand.

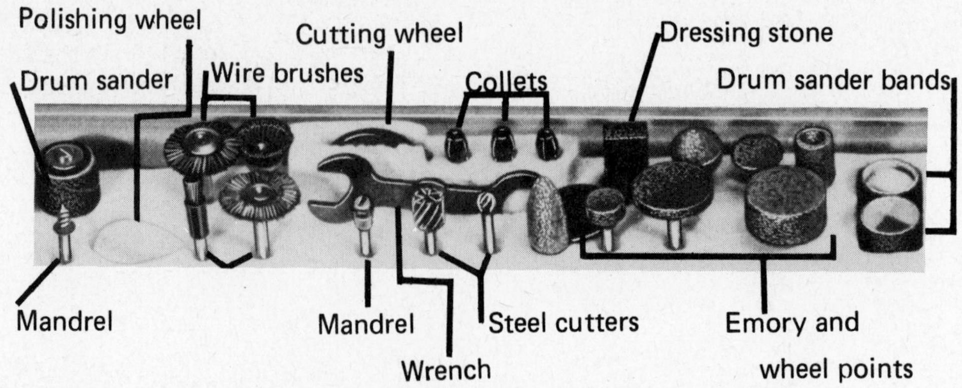

Fig. 4-13. These are some of the accessories available for the modeller's power tool: (A) drum sander; (B) polishing wheel; (C) wire brushes; (D) wrench; (E) cutting wheel; (F) collets; (G) dressing stone; (H) drum sander bands; (I) emery and silicon wheel points; (J) steel cutters; (K) mandrel

press stand (Figure 4-10) is useful when drilling accurate holes. The motor can be raised or lowered on the column to accommodate workpieces of varying sizes, but it is held in a fixed position when in use. The *table* may be raised or lowered by means of another adjustment if the operator wants to drill a hole to a specific depth. A depth lock is also provided. The universal stand (Figure 4-11) is useful for holding the tool during grinding, buffing and polishing operations. An adjustable wing nut lets you set the tool in a convenient position. A mounting hole is provided for mounting the stand on a bench.

The router attachment (Figure 4-12) allows you to use the tool as a miniature router. An adjuster on the attachment sets the depth of cut. The router can be used freehand or with a guide that is run along the edge of the workpiece. The guide enables you to make perfectly straight cuts.

Since modeller's power tools function at high speed rather than high torque, you should apply only light pressure against the workpiece. In this way, you not only avoid unnecessary tool marks and overheating, but also maintain full control of the tool.

Never stall the motor or slow it down near the stalling point. A sudden stall could over-

Fig. 4-14. Shaping a surface of a wood model.

Fig. 4-15. Sanding.

Table 4-3. Typical uses of modeller's power tool accessories

Accessory	*Typical Uses*
High-speed steel cutters	Cutting grooves and countersinking in soft metals. Internal carving of plastic. Routing in wood. Carving and hollowing soft and hard woods. Slotting and grooving woods and plastics.
Tungsten carbide cutters	Removing flash or unwanted details from cast metal models. Engraving in metal dies, shop and garden tools.
Small engraving cutters	Adding detail to ceramics, wood carvings, and jewelry castings or settings. Working on printed circuit boards, bakelite and soft metals.
Emery wheel points	Sharpening. Grinding flash from castings and molds. Smoothing welded joints. Deburring. Shaping.
Silicon grinding points	Engraving and drilling holes in glass, ceramic, and other hard, brittle materials. Grinding and shaping hard steels.
Polishing wheels	Polishing metal surfaces. Removing surface scratches and polishing plastics. Polishing jewelry.
Drum and disc sanders	Rough shaping, sanding and smoothing.
Wire brushes	Removing rust and corrosion. Polishing cast or machined metal surfaces.
Bristle brushes	Cleaning and polishing.
Cutting wheels	Cutting rods, tubing, light bar stock. Cuts hose clamps and frozen bolts.
Steel saws	Slotting or cutting wood, plastic or soft metals. Cutting thin-shell fiberglass molds.
Router bits	For routing, inlaying and mortising in wood and other soft materials.

Fig. 4-16. Carving.

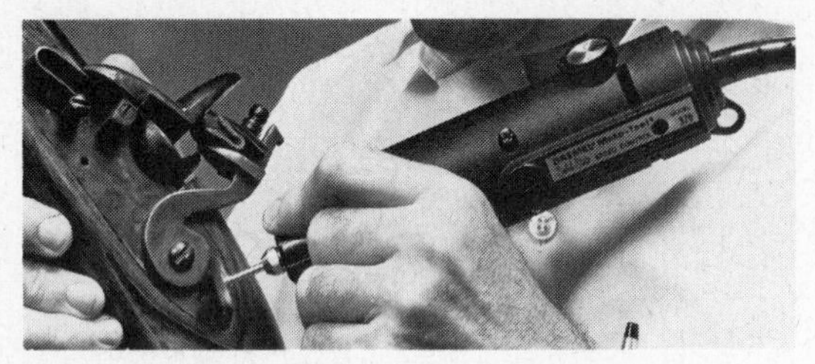

Fig. 4-17. Polishing.

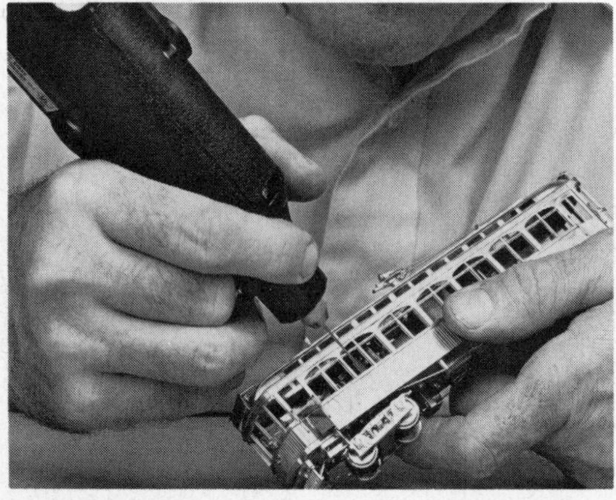

Fig. 4-18. Removing unwanted details from a cast model.

load the motor and, on the flexible-shaft model, possibly break or kink the flexible shaft.

Modeller's power tools are so small that you may overlook safety. But remember, the shafts of these tools are rotating at up to 35,000 RPM. Handle them with care. Wear safety glasses or goggles at all times when grinding or routing.

Figure 4-13 illustrates some of the accessories of the modeller's power tool. Typical uses of the various types of accessories are described in Table 4-3 and are shown in Figures 4-14 through 4-18.

4-8. Planer

A planer is used for cutting rabbets or for rough-planing flat surfaces prior to finish sanding. High-speed tungsten steel cutters turning at up to 25,000 RPM are used on wood, plastic, composition and aluminum workpieces. A planer makes cuts from paper thin up to 3/16″ on stock as wide as 2″ to 2-13/16″. Planer motors range up to 1/2 HP.

Three types of planes are available: *block planes, full-size portable planes* and *plane attachments to router motors.* The block plane shown in Figure 4-1f planes up to 1-13/16″ wide (a dressed 2″ × 4″ edge thickness) and from paper thin to a depth of 1/64″. It fits into the palm of the hand and is especially useful in planing end grains. A removable steel fence is used to guide the plane for 90° cuts; a bevel planing fence attachment allows accurate bevel planing from 0° to 45° angles.

The full-size plane (16″) can cut stock up to 2-13/16″ wide and 3/32″ deep on a single pass. Chip disposal is provided. An adjustable rip fence allows bevel cutting from -15° to +45° angles.

A plane attachment for a router allows you to get double duty from your router. If you have only an occasional need for a plane, you should consider this approach because the planer is probably the least needed power hand tool for the home shop. The planer attachment has a 60° bevel planing range (-15° to +45°), chip disposal, depth of cut adjuster and trigger switch control.

To use the router, first secure the workpiece in a vise or other clamping tool. Then, before plugging the electrical cord in, take hold of the plane with two hands – one on the trigger switch handle and the other on the motor or forward knob. The block plane can be used with one hand, but two hands give you better control. Set the depth for the desired cut; it's better to take a number of shallow cuts than to take one deep cut. The first cut, however, should always be shallow until you get the feel of the cutting action on your workpiece. Loosen the fence bevel adjustment and set

Fig. 4-19. Using a forward pressure and a sideward pressure, make a continuous stroke from one end of the workpiece to the other end.

the fence to the desired angle of bevel. Tighten the bevel adjustment. Now you're ready to apply power.

Set the front of the planer on the edge of the workpiece with the cutters remaining off the workpiece. Press the trigger switch and let the motor come up to full speed before beginning to cut. Hold the planer parallel to the edge of the workpiece and the fence firmly against the workpiece (Figure 4-19). Using forward and sideward pressure, make a continuous stroke from one end of the workpiece to the other end. At the end of the stroke, lift the planer off the workpiece and return it to the starting position for the second and subsequent strokes.

Planer blades are quickly and easily removed for replacement and sharpening. Take the blades to a professional for sharpening. Guides for setting the blade are included with the planer.

4-9. Reciprocating Saw

The reciprocating saw is a heavy-duty, all-purpose saw similar to the sabre saw (Section 4-11) except that the blade movement is horizontal rather than vertical. Blades up to 12″ long can be installed in the reciprocating saw for pruning trees and shrubs, cutting fence posts and logs up to 13″ in diameter, bar stock, steel pipe, metal conduit, wood, plastic and composition board. The saw is ideal for pocket cuts, roughing in, scrollwork and rip or cross sawing. In structural or remodelling work, it can be used to cut completely through a wall.

The removable 3″ to 12″ blades of the reciprocating saw stroke from 3/4″ to 1-1/4″. The blades are driven by a 1/5 HP motor. Motor speeds are either two-speed (1600 strokes per minute for steel, aluminum and brass, or 2000 SPM for wood, plastic and composition) or variable from 0 to 2400 SPM. A removable handle is available for additional control on large cutting jobs.

As you can gather from this description of the uses of the reciprocating saw, it is of little use to the home owner, but is a real workhorse for the carpenter, electrician, plumber or remodeller.

A multiposition foot at the front of the saw can be set in three different positions for use in flush cutting, rip and crosscutting or scrollwork. The operator selects the speed of the saw with a variable control located at the trigger on-off switch. Blades are removed and replaced with an Allen setscrew wrench. Use the shortest blade possible for the job.

To saw with the reciprocating saw, place the foot firmly against the workpiece. Set the saw to the desired speed range. Hold it with both hands and squeeze the trigger on-off switch. The blade moves back and forth, cutting on the backstroke.

When cutting holes for pipes, bore a hole first to allow the blade to pass through. Plunge cuts can be made when short blades are in the saw by firmly holding the lower part of the foot (the part below the blade teeth) on the workpiece with the blade point stretched out ahead. Start the motor (at a slow speed) and begin the cut. As a groove is made, gradually raise the saw until the blade penetrates the workpiece.

WARNING: Be sure that there are no electrical wires, conduits, water lines or gas lines in the path of the blade.

Figure 4-20 illustrates some of the blades available for reciprocating saws. The number

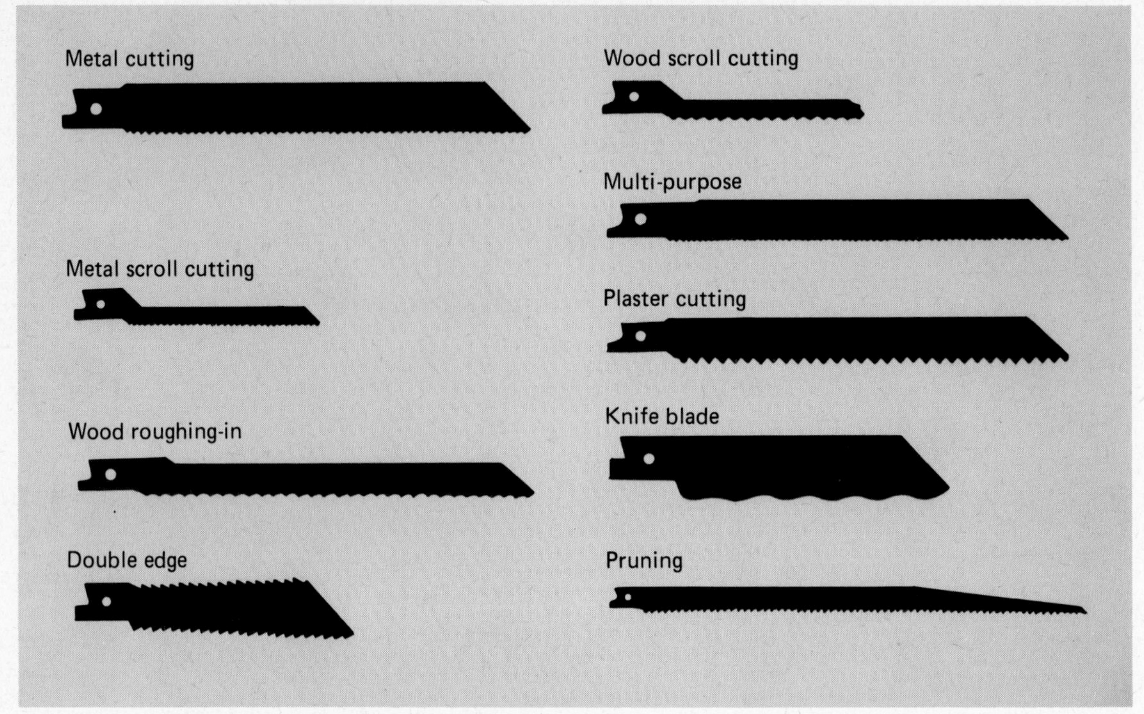

Fig. 4-20. A typical selection of reciprocating saw blades.

of teeth per inch ranges from 3-1/2 to 32; blade lengths range from 3-1/4″ to 12″. Remember that the fewer the teeth per inch, the rougher the cut. Blades with 3-1/2 to 6 teeth per inch are used for sawing wood; 6 to 10 teeth per inch for general-purpose sawing; and 10 to 18 teeth per inch for metal cutting.

4-10. Router

The router is the hand power tool with which you can show off your most creative workmanship; it's fun to use and it can produce artistry in wood, plastics and compositions. It doesn't have many practical applications for the home owner or do-it-yourselfer, but it does have a number of worthwhile applications for the carpenter.

The router is used by the artistic craftsman to produce intricate contours in wood, multi-curved moldings and edges, relief panels, trim work on cabinets and bookshelves, sign engravings and delicate grooves for intricate inlays. The serious, advanced craftsman and do-it-yourselfer use the router to make rabbet and dovetail joints, dadoes, grooves and bead moldings. The carpenter uses the router on a production basis to plane edges, mortise doors for hinges, and trim counter topping.

Incidentally, buying just a router doesn't buy you much. In addition, you need a good selection of bits and often an additional attachment for your particular job. Naturally the cost of the router mounts as you add bits, straight and curved guides, power plane, hinge mortising template kit, dovetail kit, slot- and circle-cutting attachments and veneer-trimming guide. It can be an expensive tool. Thus, if you have need to use the router only occasionally, you should borrow one from a friend or rent one. The router, by the way, seems to be one of those power tools that's often sold on sale.

When selecting a router, look for a construction that utilizes bearings and has a large high-speed horsepower motor; the greater the work load, the greater the horsepower needed. Motor sizes from 3/4 to 1-1/2 HP and speeds of 18,000 to 35,000 RPM are available. A 3/4 HP motor at 22,000 RPM is sufficient for home use, whereas a 1 to 1-1/2 HP at 22,000 to 25,000 RPM is used commercially.

Before inserting any bits or making any adjustments to the router, be sure the electrical cord is unplugged. Then, following the manufacturer's instructions, secure your bit into the router collet. Because it rotates at such high speed, it is extremely important that the bit be installed securely. Most routers use either a locknut and a collet nut to secure the bit or have a locking device to keep the shaft from rotating while the collet nut is being tightened.

Next (with the electrical plug still disconnected), set the depth of cut. Calibrations are usually in 1/64″ increments from 0 to 1-1/2″. Place the router on a flat surface and loosen the height lock knob. Turn the height-adjust-

Fig. 4-21. Use a piece of straight scrap wood or a template to guide the router.

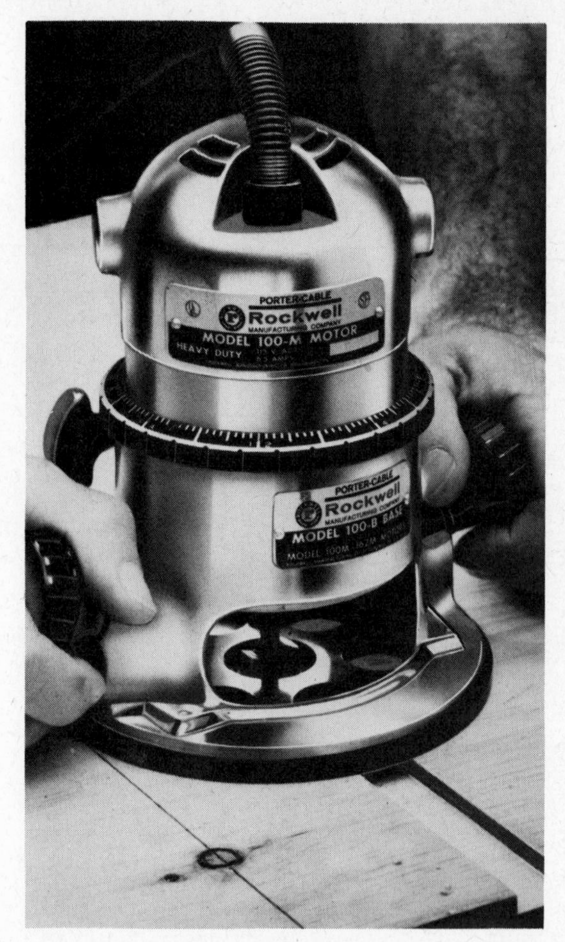

Fig. 4-22. Grip the router firmly at all times. Here, a dado is being cut with a mortise bit.

ing knob until the bit just touches the surface. Now, position the bit so that it can be projected below the base to the desired setting. Rotate the height-adjusting knob to set the bit into the desired position. Tighten the locknut.

If the router has a power off-on switch, it will be located on the motor; if it has a trigger switch, it will be in one of the guide handles. Where trigger handles are utilized, a locking button may be incorporated to hold the trigger switch on. Further pressure and a releasing of the trigger unlocks the spring-loaded lock.

Examine the workpiece and plan how you will use the router. If the job requires working more than one piece, plan to do the same cutting on similar workpieces consecutively. Secure the workpiece firmly with a clamping

Fig. 4-23. The average cut should not exceed 1/4 inch depth per pass. Here, a Roman ogee bit is used to cut a decorative edge.

device such as a vise, clamp or jib. Remember that the clamping device must be thinner than the workpiece so that the router can pass over it.

Unless you are planning to perform freehand routing, clamp a guide (Figure 4-21) or template to the workpiece. Attach any accessories that you need.

Make sure that a sharp bit is secured in the router before connecting the electrical plug into power. Grip the router firmly at all times and especially during turn-on because of the initial starting torque (Figure 4-22). Turn the switch on and let the router motor come up to full speed before you begin cutting. Hold the router against the guide and let the bit cut into the workpiece, feeding left to right as the router is moved across the workpiece. Feed the router at a moderate speed – too slow a feed may burn the wood and heat up the bit, causing a loss of temper in the bit metal; too fast a movement slows the motor and causes it to overheat. Keep both the motor and the bit from overheating. If the motor slows down too much, decrease the depth of cut and make two or more passes.

The average depth of cut should not exceed 1/4″ per pass (Figure 4-23) when large cutters are used. In cutting hardwoods, make two or three passes – that is always more advisable than risking possible damage to the motor or cutter by making one deep pass.

When making straight cuts, keep the router against the clamped-down guide. The edge guides that come with the router can be used for cuts that are close to the edge. In cutting circles or arcs, the edge guide is pinned down.

Freehand cuts are used for special designs and for inlay work. Mark the pattern first. On inlay work, set the router depth so that the depth of cut is slightly thinner than the inlay material.

Practise first on a scrap of wood of the same species as the workpiece. This gives you the feel of the router on the workpiece material (Figure 4-24).

The following accessories are used with the router: bits (Figure 4-25), dovetail template, butt hinge template, laminated plastic trimmer attachment, alphabet and numbers template, pantograph template (the original size is traced with a stylus – one-half size is made) and a router table that is used to convert the router into a shaper.

Two types of cutter bits are available: a one-piece or a screw-on bit that allows

Fig. 4-24. Practice on scrap wood first. Here a bead is being cut.

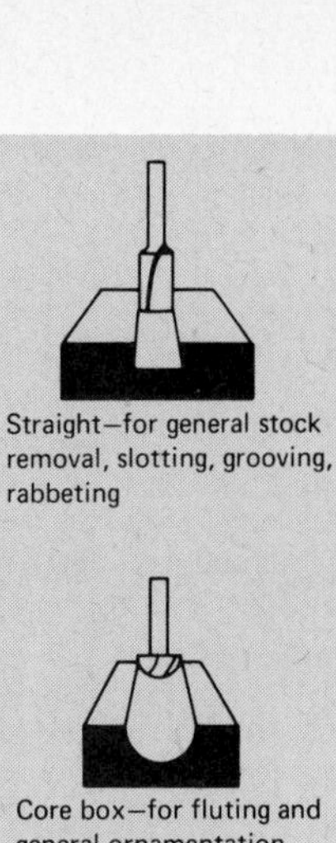

Straight—for general stock removal, slotting, grooving, rabbeting

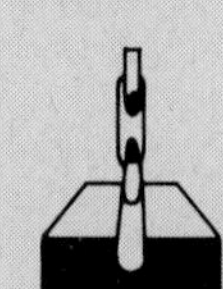

Veining—for decorative free-hand routing such as carving, inlay work

Sash bead—for beading inner side of window frames

Sash cope—for coping window rails to match bead cut

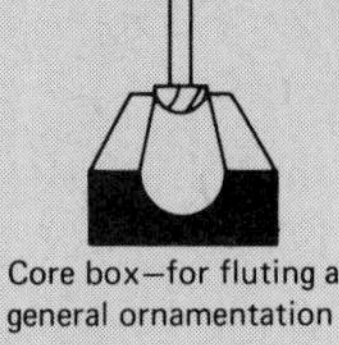

Core box—for fluting and general ornamentation

Dovetail—for dovetailing joints. Use with dovetail templet

Corner round—for edge rounding

Bead—for decorative edging

Cove—for cutting coves

45° bevel chamfer—for bevel cutting

Mortise—for stock removal, dados, rabbets, hinge butt mortising

Rabbeting—for rabbeting or step-cutting edges

Roman ogee—for decorative edging

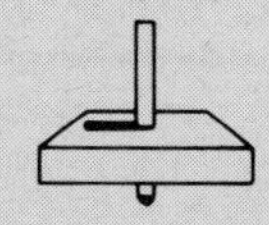

Panel pilot—for cutting openings and for through-cutting

Pilot spiral (down)—for operations where plunge cutting is required in conjunction with templet routing, using the pilot guide

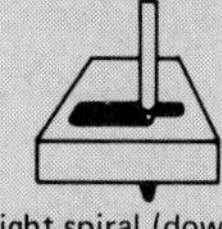

Straight spiral (down)—for through cutting plastics and non-ferrous metals; also for deep slotting operations in wood

Straight spiral (up)—for slotting and mortising operations particularly in non-ferrous metals such as aluminum door jambs

V-groove—for simulating plank construction

Spiral—for outside and inside curve cutting

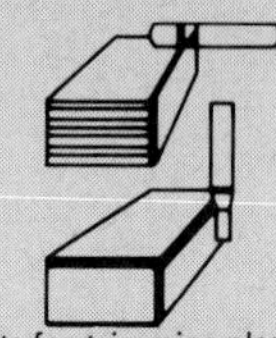

Bits for trimming plastic laminates. Solid carbide or carbide-tipped bits for flush and bevel trimming operations Solid carbide self-pivoting flush and bevel trimming bits

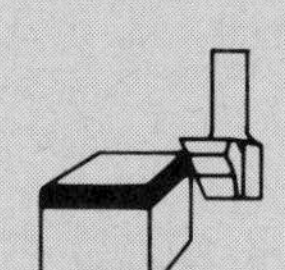

Solid-carbide combination flush/bevel trimming bit

Carbide-tipped bevel trimmer bit

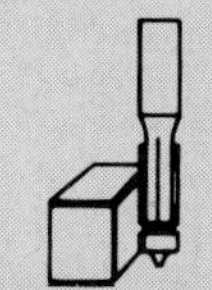

Carbide-tipped ball bearing flush trimming bit

Carbide-tipped 25° bevel trimmer kits

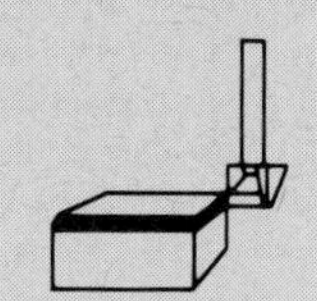

Carbide-tipped combination flush/bevel trimming bit

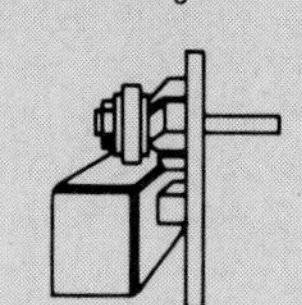

Carbide-tipped 15° backspiash trimmer—with $\frac{5}{16}$'' diameter hole

Fig. 4-25. Router bits.

different cutters to be mounted on one shaft. Only high-speed steel cutters or carbide-tipped cutters should be purchased; other types just do not hold up. Carbide-tipped cutters are for use on laminates, plastics and plywood.

Keep cutters sharp and clean. During operation, keep the cutters and the motor from overheating and smoking.

To sharpen a bit, grind only the inside of the cutting edge. Be sure to grind the clearance angle to the same size as the original. Never grind the outside diameter of the bits because they are specially ground for proper clearance. If you're not experienced in sharpening, you should either have a professional regrind your bits or purchase new ones when required.

Before and after using the router, inspect the bit shafts for burrs and remove them with oilstone. Pitch and gum can be removed with kerosene. Keep a light coat of machine oil on the bits when storing them and be sure they do not come into contact with each other or with other tools.

Occasionally clean and wax the base of the router so that there is a minimum of friction between the base and the workpiece during operation.

4-11. Sabre Saw

The sabre saw, also called a jigsaw, is a reciprocating saw used to cut thin-gauge metals, bars, angles, fiberglass, plaster, tile, wood, plastic, composition and pipe. It is especially useful in cutting square, circular, irregular or intricate cutouts in sheet materials, walls and air ducts. Approximate capacities are: softwood, 1-5/8″; hardwood, 1″; aluminum, 1/4″; and mild steel, 1/4″.

The sabre saw is available with three types of speed control: fixed, two-speed and variable. The fixed-speed saw has a stroke per minute (SPM) of about 3000. The two-speed saw has an SPM of about 3200 for the high speed and 2700 for the low speed. Speed selection is controlled by a two-position switch located on the handle. The variable-speed saw has a speed range from 0 to approximately 3500 SPM and is controlled by a trigger-type switch. Motor ratings are from 1/4 to 1/2 HP.

The two-speed and variable-speed sabre saws have been developed to increase the range of applications of the sabre saw. The high speeds are used with wood and compositions while the lower speeds are used for metal and plastic. The variable speed lets you select the proper speed for the job (the harder the material, the slower the speed); it also lets you slow down for cutting curves.

The main items on the sabre saw are a case, a trigger switch, a blade chuck, a saw blade, a shoe plate and a sawdust blower (not available on all models).

Most sabre saw blades are fixed in one plane of operation with the blade moving up and down; to turn the saw blade around, the complete saw has to be turned. One manufacturer, however, has a blade that can be rotated without turning the saw itself or without removing power from the saw. To turn the blade, pressure is applied in a downward direction to the top handle; the blade can be locked in a straight position in any one of four 90° angle positions.

Fig. 4-26. The tilting shoe is set for making bevel cuts.

	Blade type	Description of blade and use	Type of cut	Speed of cut	Blade length	Teeth per inch
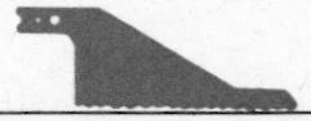	Flush cutting	Hard or soft wood over $\frac{1}{4}$" thick.	Rough	Fast	3"	7
	Plaster cutting	Special V-tooth design provides constant abrading action which is most effective in cutting plaster, masonry and high density plastics.	Rough	Fast	$3\frac{5}{8}$"	9
	Double cutting	Most wood and fiber materials. Tooth design allows for cutting in both directions with equal speed.	Rough	Fast	3"	7
	Double cutting	Cuts most wood and fiber materials. Tooth design allows for cutting in both directions with equal speed and quality of cut.	Medium	Medium	3"	10
	Skip tooth	Cuts most plastics and plywood. Special tooth design with extra large gullets provide extra chip clearance necessary for cutting plywood and plastic.	Rough	Fast	3"	5
	Wood cutting coarse	Cuts soft woods $\frac{3}{4}$" and thicker. Canted shank provides built-in blade relief, thus helping to clear the saw dust and cool the blade.	Rough	Fastest	3"	7
	Wood cutting fine	Cuts soft woods under $\frac{3}{4}$" thick. Canted shank provides built-in blade relief, thus helping to clear the saw dust and cool blade. More teeth per inch allows for finer quality of cut.	Medium	Medium	3"	10
	Wood cutting hollow ground	Hard woods under $\frac{3}{4}$" thick. Hollow grinding provides no tooth projection beyond body of blade, thus imparting an absolutely smooth finish. Canted shank for blade clearance.	Smooth	Medium	3"	7
	Metal cutting	For cutting ferrous (iron) metals $\frac{1}{16}$" to $\frac{3}{8}$" thick and nonferrous (aluminum, copper, etc.) $\frac{1}{8}$" to $\frac{1}{4}$" thick.	Medium	Medium	3"	14 to 32
	Hollow ground	For cutting plywood and finish materials $\frac{1}{2}$" and thicker where fine finish is desirable. Hollow ground for very smooth finish on all wood products. Provides the longest life woodcutting blade possible.	Fine to medium	Medium	$4\frac{1}{4}$"	6 to 10
	Knife blade	For cutting leather, rubber, composition tile, cardboard, etc.	Smooth	Fast	3"	Knife edge
	Fleam ground	For cutting green or wet woods $\frac{1}{4}$" to $1\frac{1}{2}$" thick. Fleam ground provides shredding type cutting action which is most effective in sawing hard, green or wet materials. Provides longest cutting life possible.	Smooth to coarse	Medium	4"	10
	Scroll cut	For cutting wood, plastic and plywood $\frac{1}{4}$" to 1" thick. Set teeth and thin construction allows this blade to make intricate cuts and circles with radii as small as $\frac{1}{8}$"	Smooth	Medium	$2\frac{1}{2}$"	10
	Wood cutting coarse	Cuts most plastics and wood up to 4" thick. Special tooth design with extra large gullets provide extra chip clearance for fast cutting in thicker materials.	Rough	Fast	6"	3

Fig. 4-27. Types of sabre saw blades.

The cutting action of the sabre saw is upward toward the bottom of the shoe plate. This upward cutting action pulls the shoe plate tightly to the workpiece during cutting and relaxes its pull when the blade is moving away from the shoe plate. Therefore, it is important that the workpiece be placed with its good side down; the feathered and splintered edges will then appear on the rough side of the workpiece.

The reciprocating action of the blade causes the saw to vibrate strongly during operation. Some of the vibration is eliminated by clamping the workpiece and applying firm, steady, downward pressure during the sawing. The calibrated shoe plate is adjustable with respect to the blade through a range of 0 to 45° left or right for cutting bevels. A single screw located on the angle protractor is used to adjust and lock the shoe plate to the required angle (Figure 4-26).

Sabre saw blades are available in a wide range of styles to meet specific applications. The determining criteria in blade selection are the material to be cut and the type of finish to be left by the blade. The finish or type of cut is classified as rough, medium and fine. Narrow blades are used for curved work and wide blades for straight work. Choose the shortest blade possible for the job.

Blade lengths are from 3″ to 4-1/4″ long. The longer blades are usually used for cutting wood. Two forms of blade teeth are available for most general work, the stagger-tooth blade and the wavy-tooth blade. The stagger-tooth blade has teeth that are set in an alternating pattern of right, left, right, left. This blade is most often used on soft materials such as wood, plastics and aluminum. The wavy-tooth blade is used for cutting ferrous metals of up to 1/4″ thick and non-ferrous metals up to 1/8″ thick. Blades vary in number of teeth from 3 teeth per inch (used in cutting soft materials) to 32 teeth per inch (used in cutting ferrous metal). Figure 4-27 is included as an aid to selecting the correct blade for the material to be cut.

The blade chuck is a slotted collar located at the end of the output shaft of the saw. Generally, there are two screws (or hex setscrews) located on this collar, one in the front and the other on the side, 90° from the front. These screws align and lock the blade to the output shaft.

To install a blade in the sabre saw, first disconnect the power plug from the outlet. Then loosen both screws in the blade chuck until the shank of the blade can be inserted and bottomed. With the blade bottomed, turn the front screw until it just touches the blade. Next, tighten the side screw firmly against the blade. Then tighten the front screw securely.

To make a cut with the sabre saw, hold the saw firmly in one hand and place the shoe plate firmly on the workpiece with the blade off the workpiece. Start the saw and let the motor run up to speed. Move the saw onto the workpiece and begin cutting (Figure 4-28). Feed the blade into the workpiece at a moderate feed and speed; don't force the saw – let the blade do the cutting.

To saw thin metal, you should either use a backup piece of plywood or masonite behind the metal or sandwich the metal between two pieces and saw through the entire sandwich. The backup piece aids in cutting clean edges on the workpiece and it also helps to dampen vibration, the biggest source of blade breakage.

In sawing thick metal workpieces, such as 1/8″ or 1/4″ sections, ample support should be supplied. If sawhorses or some other supports are used, always keep the part of the

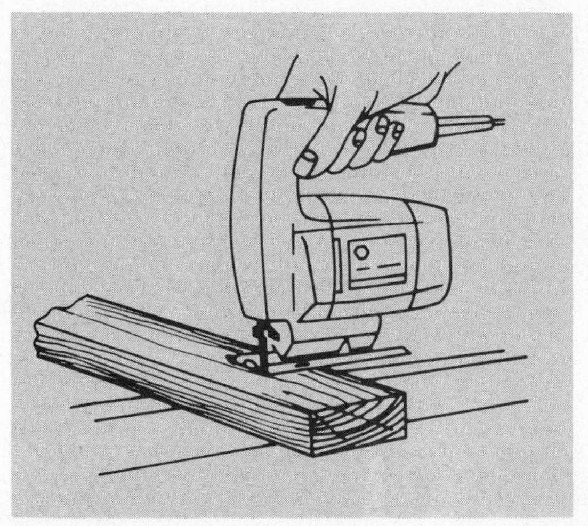

Fig. 4-28. Hold the sabre saw firmly.

Fig. 4-29. Making a plunge cut.

workpiece being cut as close as possible to the support to eliminate vibration.

The sabre saw can make plunge cuts into a workpiece. A plunge cut is a cut through some inner part of the workpiece without cutting in from the edge and without drilling a hole for the blade. To make a plunge cut, angle the sabre saw onto the forward edge of the shoe plate (Figure 4-29) with the teeth to the surface. Turn the motor on at a slow speed and begin cutting. As the blade cuts a groove, slowly lower the back of the saw until the blade finally plunges through the surface.

The rip fence and circle guide are two accessories used with the sabre saw to make straight cuts using a reference edge and to cut circles. The rip fence and circle guide insert into the slot in the shoe of the sabre saw and can be locked at any setting. The rip fence guides the sabre saw in cutting strips parallel to a workpiece edge (Figure 4-30).

To cut circles, the combination rip fence and circle guide is removed and reinstalled with the crossbar edge up. The required radius (the distance between the pivot hole and the blade) is then set up and the position locked. By placing a pin through the crossbar into the center of the required circle, the blade can be inserted into a predrilled hole or slot and the circle cut out using the pin as a pivot point. Sabre saw blades are expendable and should be replaced when they become dull.

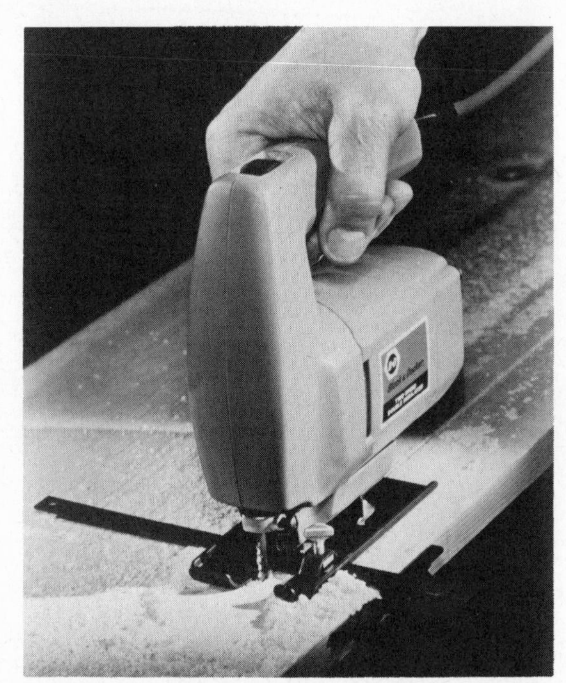

Fig. 4-30. Use the rip fence guide for making cuts parallel to the edge of the workpiece.

4-12. Sanders (Belt, Disc, Finishing)

If there is an unpleasant job in the shop, it must be sanding. It's the sanding, however, that turns the workpiece into a finished, seemingly professional product. The three sanders described in this section will aid you in finishing a workpiece and will save you time and energy.

The belt sander is used for removing paint, stains and varnishes, for sanding rough lumber, for rounding edges and for smoothing badly scarred surfaces. It is used where a lot of material is to be removed with speed. It can sand flush to vertical surfaces.

The belt sander uses a continuous abrasive-coated belt that rides over two drums and a sanding plate. The rear drum is powered, whereas the front drum is an idler that is adjustable for tension and for true tracking of the belts across the sanding plate (4″ × 6″)

and drums. A dust pick-up bag is provided. The belts are driven by motors of 3/4 to 1-1/2 HP at approximately 1200 to 1500 feet per minute.

The size of the belt sander is determined by the belt width. Most home belt sanders are 3″ wide.

Belts may be changed rapidly either by activating a releasing lever or by pushing the forward drum toward the rear. Noting the direction of travel marked on each belt, carefully mount the belt on the drums. Press the drum release and let the front idler drum go back into place. Adjust the travel knob to alter the angle of the front drum so that the belt tracks perfectly. Make several momentary power on-off tests to check the trueness of the belt on the drums. The final trueness adjustment can be made with the belt running.

Secure the workpiece. Grip the belt sander firmly with two hands and turn it on. Let the motor get up to speed before making contact with the workpiece. Place the sander to the workpiece and begin sanding. Move the sander continuously to avoid removing excessive material from any one area. To cut rapidly, move the sander at a 45° angle across the workpiece. Keep the sander flat at all times and only bear down hard enough to cut. Make long, straight, continuous sanding strokes. Do not use too much pressure. Do not tilt the sander as you move it off the workpiece or you'll have rounded edges.

The disc sander is excellent for removing paint and rust from boats and cars and for polishing. It is used extensively by auto body repair shops and occasionally in the home shop. It is most often used on curved surfaces.

The size of the disc sander is determined by the size of the sanding disc used, usually 5″ or 7″. The sanding discs are mounted onto a rubber or fiber backing-pad. The sander motors range from 7/8 to 1-1/2 HP and have speeds of up to 5000 RPM; two-speed models are available with speeds of approximately 2400 and 4700 RPM. Polishing is usually done at low speed.

Two types of disc sanders are available. One is an in-line model that looks like a drill; the other is an offset model (Figure 4-31). The offset model is handier than the in-line.

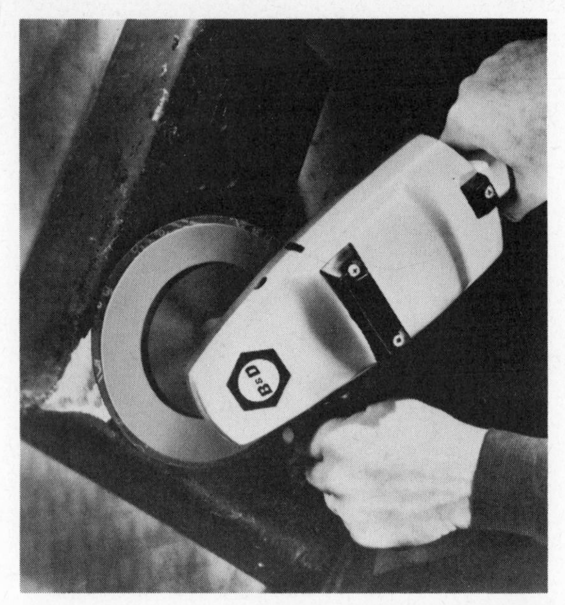

Fig. 4-31. Tilt the disc sander slightly against the workpiece.

Apply power to the disc sander and let it reach maximum speed before touching it to the workpiece. Hold the sander with two hands (Figure 4-31) and tilt the disc slightly against the workpiece. Apply just enough pressure to bend the rubber backing. The main disadvantage of the disc sander is that it digs into the workpiece if not controlled and kept moving. It does not give as good a finish to a workpiece as the finishing and belt sanders.

The finishing sander is used for fine finish sanding of wood, metal and plastic. With the proper abrasive papers (Section 1-3), it is also good for sanding rougher work. Because of its flat, square base, it can be worked directly into a corner and provides a better finished sanding job than can be done by hand.

There are three types of finishing sanders: *orbital, straight-line* and a combination *orbital/straight-line*. The orbital/straight-line sander costs a little more than the other two, but is well worth the price. The rectangular base of the orbital finishing sander, called a *platen*, moves in an orbital path of about 3/16″ diameter at about 4000 orbits per minute (OPM). This orbital action is for fast, tough sanding, removal of old paint varnish, and for initial final sanding operations. Straight-line sanders produce finishes so fine they appear to have been done by hand. They operate at approximately 14,000 SPM and have motors of about 1/6 HP.

The combination orbital/straight-line sander provides both types of finishing sander actions by converting the orbital motion of the platen into a back-and-forth, straight-line motion. The operator selects the action by moving a lever (or rotating a screw-headed cam) located between the motor housing and the platen.

The platen is lined with a rubber or felt face to add a firm, flexible backing to the sanding sheets during use. The sanding platens are made in two popular sizes: 3-5/8″ × 9″ and 4-1/2″ × 11″. The smaller dimension of each of these two sizes is the width of the sanding paper, but the longer dimension includes the extra length needed for clamping the sheet at each end of the sander platen.

There are several ways to attach the sanding sheet to the platen. Some sanders employ a spring-clamping device, whereby one end of the long length of the sheet is rolled around the end of the platen and clamped. The second clamp is employed after stretching the sheet taut over the platen.

Fig. 4-32. The finishing sander can be purchased with a combination orbital/straight-line action.

Fig. 4-33. This small finishing sander is useful on curved surfaces.

A second method uses a split-ratcheted cylinder located at each end of the pad. To secure a sanding sheet to the pad, insert one end of the sanding sheet into the slotted front cylinder (opposite the handle). Draw the opposite end of the sheet taut across the platen and insert its end into the slotted ratchet cylinder. Using a screwdriver or the tool supplied with the sander, tighten this cylinder until the sanding sheet is taut.

After you select the type of action (orbital or straight-line) you desire, install the abrasive paper onto the sander as previously described. (If you are just beginning to sand a workpiece, the orbital action should be selected first, followed by straight-line). Plug it into the power socket, and holding it in one hand, turn it on. Now, place the sander onto the workpiece and move the sander with the grain of the wood. Although the sander can be held in one hand (Figure 4-32), it is better to use two hands to guide it. Do not apply pressure with the hands; let the weight of the machine provide the pressure.

The soft, felt platen is adequate for most work. However, on wood workpieces that have both soft and hard grain in them (such as fir), it is recommended that a piece of hardboard be placed between the sandpaper and the felt pad; this will permit all of the grain to be cut at the same level rather than the softer grain being cut faster and hence deeper than the harder grain.

Keep abrasive papers clean. Remove dust from them by rubbing the paper with a stiff brush. Brush and vacuum the exposed surfaces of sanders on a regular basis. Wipe the exposed surfaces with a rag dampened with turpentine.

4-13. Soldering Irons

Soldering irons are used to solder metal pieces together. You'll probably use the soldering iron most frequently to repair electrical connections in appliances and to solder wire and components in electronic kits you may construct. Soldering irons usually take one of four forms: miniature soldering iron, pistol-grip soldering iron, instant-heat soldering iron (Figure 4-1m, n, and p, respectively) and as an attachment to the propane torch (Section 1-12).

The miniature soldering iron is used almost exclusively in the electronics industry for soldering wires, transistors, integrated circuits, printed circuits and other electronic components. It is available with interchangeable thread-in heating elements from 10 watts to 50 watts. Tips are also interchangeable thread-in units and are available in a number of shapes, including chisel, offset chisel, pyramid, pencil, long-taper chisel and spade. The miniature iron is plugged in and left on while it is in use.

The pistol-grip iron features a 75-watt sealed nichrome element. The soldering tip, made of copper, is easily replaced by removing two hex nuts. A spotlight on the housing lights up the work area. The iron is for heavy-duty wiring and should be plugged in and left on when in use.

Table 1-6 lists the various solders and solder fluxes (flux is a substance used to promote fusion of the solder and the wires or components) used for different soldering jobs. A multi-core, resin-type solder is always used with electronic equipment; never use acid-core solder because it will eventually eat away the component leads.

Upon receiving a new soldering iron or tip, the first thing you must do is *tin* the tip. You should follow the manufacturer's specific instructions for accomplishing this task; basically, it involves heating the tip for a predetermined time period, applying a bit of solder to the tip, wiping the tip clean and repeating this process one or more times until the tip is tinned. A tinned tip enables the solder to flow from the heated tip to the connection being soldered.

To solder properly, you must first heat the soldering iron to its operating temperature. Then, wipe the tip across a damp sponge or cloth to clean it and apply a small amount of solder to the tip. Place the flat part of the tip against the junction or connection to be soldered. Let the tip heat the junction. Apply solder to the junction and let the heat of the junction melt the solder. When the solder flows into the junction, remove the iron and let the junction cool and solidify before moving the parts of the soldered areas. A bright soldered junction indicates a good joint. A silver-gray junction indicates a *cold soldering joint* that must be corrected. Reheat and apply a small amount of additional solder.

4-14. X-tra Tool

The X-tra tool is a multipurpose tool that operates in three ways: *drill/drive, hammer-chisel* and *hammer-drill* (Figure 4-34). With the appropriate drill, screwdriver or chisel bit installed, it can drill, drive screws, hammer-chisel, hammer-drill, scrape paint, chip putty, chisel wood, chisel floor tile, edge, channel, gouge, shape and slot.

In the drill/drive mode of operation, the X-tra tool chuck revolves at a variable speed from 0 to 850 RPM; it is also reversible and performs the same functions as the electric drill (Section 4-5). In the hammer-chisel mode, the chuck moves in and out at a rate varying from 0 to 33,750 blows per minute.

This mode is useful for chiselling. In the hammer-drill mode of operation, the X-tra tool moves in combined circular motion and an in-and-out motion. The hammer-drill mode is especially effective when using masonry bits to drill holes in masonry.

The X-tra tool is driven by a 1/3 HP motor and is double-insulated. It has ball-and-thrust bearing construction and has double-reduction gearing for added torque. The chuck takes up to 3/8″ bit shank.

Tools are inserted into the X-tra tool chuck as they are into an electric drill (Section 4-5). A key tightens the chuck. The trigger on-off switch applies power to the drill; the variable-speed control is an integral part of the trigger assembly. A trigger lock is provided to lock the tool at several different speeds.

To use the X-tra tool in the drill/drive mode, operate the tool in the same manner as described for the electric drill (Section 4-5). To change to another mode of operation, simply rotate the collar and change the bit. In the hammer-chisel mode, hold the tool in position before applying power. With power applied, move the chisel bit carefully along the workpiece to prevent splitting and the removal of excessive material. In the hammer-drill mode, hold the masonry bit to the marked hole location (it's a good idea to start a tiny hole by tapping the center of the hole with a nailhead and hammer) and turn the tool on. Press lightly against the tool as it makes the hole.

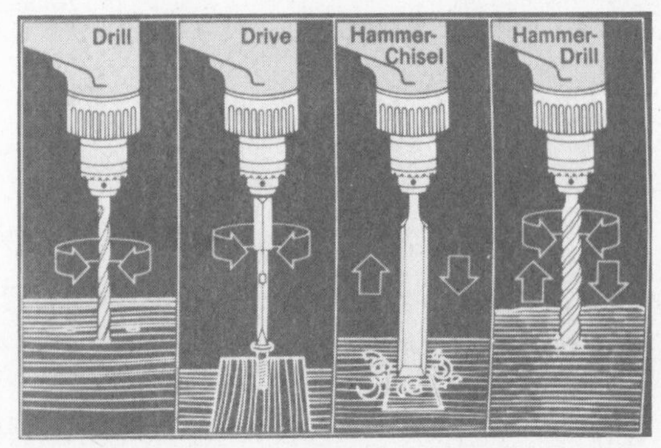

Fig. 4-34. Modes of operation of the multipurpose drill.

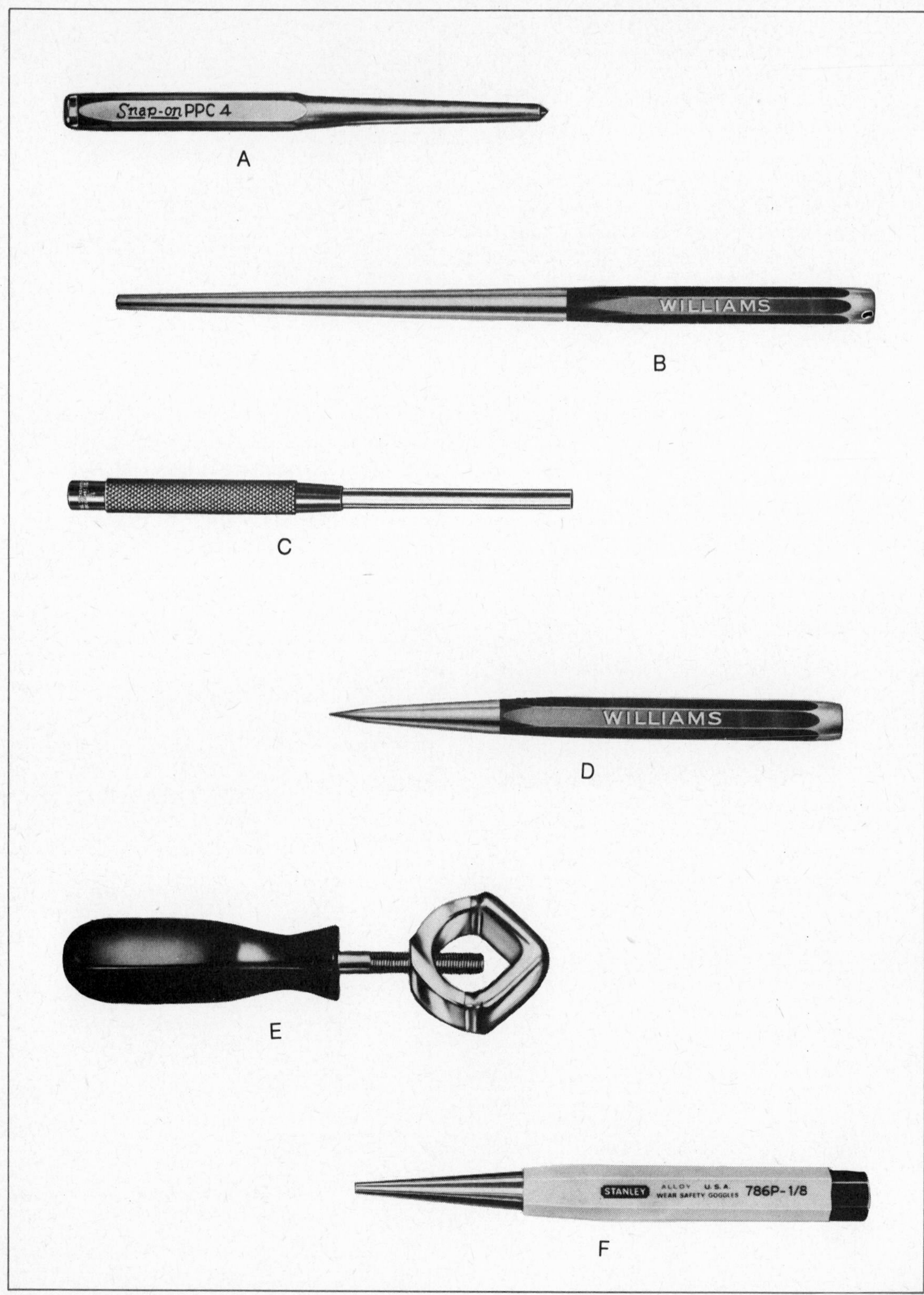

Fig. 5-1. Punches: (A) center; (B) long-taper (aligning); (C) pin; (D) prick; (E) punch-and-chisel holder; (F) solid (starter).

5 Punches

5-1. General Description

Punches are used to make indentations in metal and other materials for later drilling of holes, to mark metal for layout, to drive straight or tapered pins, and to align holes in two workpieces. The types of punches most often used in the home shop and by maintenance personnel and technicians are the *center punch* and the *solid* (starter) *punch. Pin punches* are used in applications where a large number of pins are removed. *Prick punches* are used in laying out patterns on workpieces; and *long-taper (aligning) punches* are used in heavy assembly work where two workpieces are to be aligned for assembly. A *punch-and-chisel holder* is useful for holding tools in tight spaces.

Punches are made from hardened tool steel stock. One end is shaped for the job application; the other end, called the *anvil,* is struck by the user with a ball peen hammer. You should purchase only high-quality punches with tempered anvils. The tempering keeps the anvil from mushrooming and fracturing and thereby eliminates hazardous flying splinters of metal.

5-2. Application of Punches

To use a punch, grasp it lightly in the fingers of one hand and hold it perpendicular to the workpiece. Place the point of the punch flat against the workpiece. Strike the anvil lightly with a light ball peen hammer. If additional strikes are required, check before each blow to ensure that the punch is still perpendicular and flat against the workpiece. Hold the punch in the punch-and-chisel holder when you are working in tight spaces.

Always use a punch that is the proper size. In removing pins from a hole, for example, the correct punch is the one which just fits into the hole. Hold the punch in direct line with the pin.

When working with punches, protect yourself from pieces of flying metal by wearing safety glasses or goggles. Protect others by placing a protective booth or portable screen around your work area. Repair mushroomed anvils.

5-3. Care of Punches

Long-taper, pin and solid punch points should be flat. If necessary, grind the points until they are flat and at right angles to the center line of the punch. Adjust the grinder tool rest so that the end of the punch is opposite the center of the wheel. Place the punch on the rest and rotate it as it is fed against the wheel. Keep the point cool by dipping it in water often.

Prick and center punch points are ground to 30° and 60° conical points, respectively. Adjust the grinder tool rest for the proper angle. Place the punch on the tool rest and grind as described for the long-taper punch.

5-4. Center Punch

The center punch is used to punch a conical-shaped indentation into a metallic workpiece as a starting point for a twist drill. The point of the punch, and hence the shape of the conical indentation, is 60°, which is the same as the tip of a twist drill. The indentation in the surface of the workpiece allows the operator to drill a hole without worrying that the bit will walk over the surface and cause him to drill the hole inaccurately. On workpieces where accurate layouts are required, hole centers are punched with a prick punch before the center punch is used.

The center punch is sometimes used to mark alignment points or to otherwise identify parts that have been disassembled. The center punch may be used to make indentations in very hard woods for drill point centers; in soft woods, the awl is used for this purpose.

A typical center punch is made of 1/4″ to 3/4″ hardened tool steel stock and is from 3-1/2″ to 7″ long. The punch tapers to a point of 1/8″. It is easily identified from other punches by the short, conical end which has a 60° point.

To use the center punch, hold it lightly in the fingers of one hand and slant it with the striking end (anvil) away from yourself. Holding the punch point on the layout point, raise the punch until it is perpendicular to the workpiece and firmly strike the punch anvil with a ball peen hammer to make an indentation in the surface of the workpiece.

Should you mistakenly make an inaccurate indentation, place the punch in the indentation and slant it with the point toward the true mark. Tap the anvil until the punch point and the fractured metal of the workpiece move over to this original mark. When the punch point is in line with the original mark, raise the punch to a vertical position and strike the anvil to enlarge the indentation.

5-5. Long-Taper (Aligning) Punch

The long-taper punch is used in aligning workpiece holes for installing fasteners. It is especially useful for engine installations and

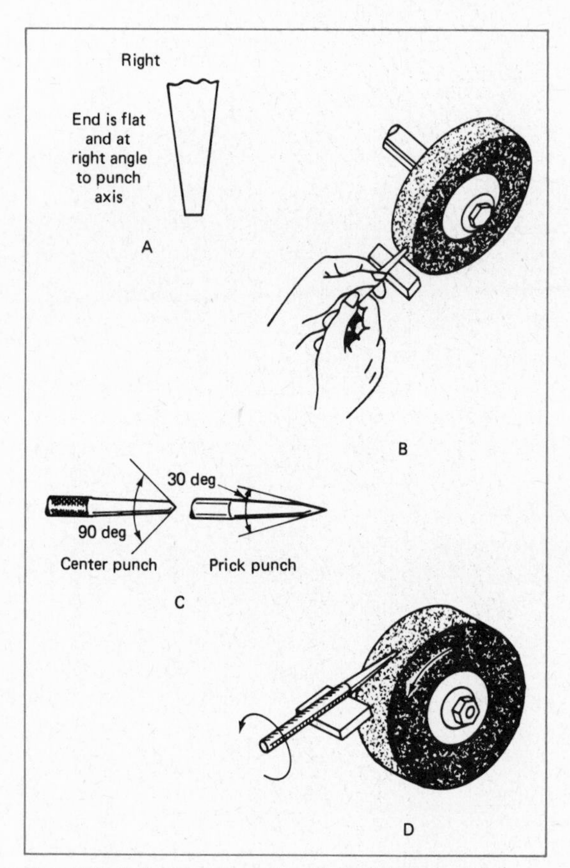

Fig. 5-2. Grinding punch points: (A) long-taper pin, and solid punch point; (B) grinding a long-taper pin, or solid punch point; (C) center punch and prick punch tip angles; (D) grinding a center or prick punch point.

replacement of springs. The tapered punch is placed through one hole where two holes in two workpieces are to line up. The workpieces are moved slightly. When the holes are aligned, the long-taper punch will drop into the aligned hole.

Long-taper punches are available in stock sizes from 3/8″ to 3/4″ with lengths of 8″ to 16″; point diameters range from 5/32″ to 3/4″.

5-6. Pin Punch

The pin punch is used to drive straight, tapered, spiral, roll, dowel, cotter and shear pins out of such objects as lathes, outboard motors, guns and rifles, or any assemblage having two pieces of metal joined by pins. It may also be used to drive bolts out of metal or wood. The pin punch is recognizable by its straight, untapered shank and flat end. Punch diameters from 3/32″ to 3/8″ and lengths from 4-1/2″ to 6-1/4″ are available. Pin punch shanks range from 1/4″ to 1/2″.

To remove a pin from a hole, use a solid (starter) punch first. Applying light blows with a light ball peen hammer, drive the pin out with the solid punch until the taper diameter of the solid punch nearly equals the diameter of the hole. Then use the fattest pin punch that will fit into the hole to complete the pin driving. In the case of tapered pins, carefully measure both ends of the hole to determine which has the smaller diameter; apply the pin punch tip to the smaller end directly against the pin. The pin punch should not be used for the first sharp blow of the hammer needed to *unfreeze* the pin because the narrow shank of the pin punch could bend or break.

5-7. Prick Punch

The prick punch is used as a starting tool in laying out metal workpieces. It is used to punch small, conical indentations in the workpiece surface to permanently mark points of a line, the intersections of lines and circles or arcs, and the centers of circles and holes. The small indentations are also used for the leg of a divider. A typical prick punch is made of 3/8″ hardened tool steel and ranges from 4-1/2″ to 6″ long; it has a long, tapered shank terminating in a conical point.

Use a prick punch in the same manner as that described for the center punch (Section 5-4).

5-8. Punch-and-Chisel Holder

The punch-and-chisel holder is used for holding a punch or chisel in hard-to-reach spaces.

To use the holder, insert the punch or chisel and tighten the handle that forces the tool against the plastic-coated head. Hold the holder and tool in one hand in position against the workpiece; tap the tool with a ball peen hammer. Before each successive tap with the hammer, check to make sure the tool and the workpiece are aligned.

5-9. Solid (Starter) Punch

The solid punch, or starter punch, is used as the starting punch in driving out straight, tapered, spiral, roll, dowel, cotter and shear pins, as well as bolts and rivets that have had the heads removed. Because of its tapered shank, the solid punch cannot be used to drive pins completely out. A pin punch should be used to drive the pin the rest of the way out. Always use the largest solid punch that fits the hole. Place the punch solidly against and in line with the pin. Tap the punch anvil with a ball peen hammer and drive the pin out until the diameter of the solid punch is approximately that of the hole. Then use the pin punch. Solid punches are made of stock from 1/4″ to 9/16″ and lengths from 4-1/2″ to 7-1/4″; point diameters are from 3/32″ to 1/2″.

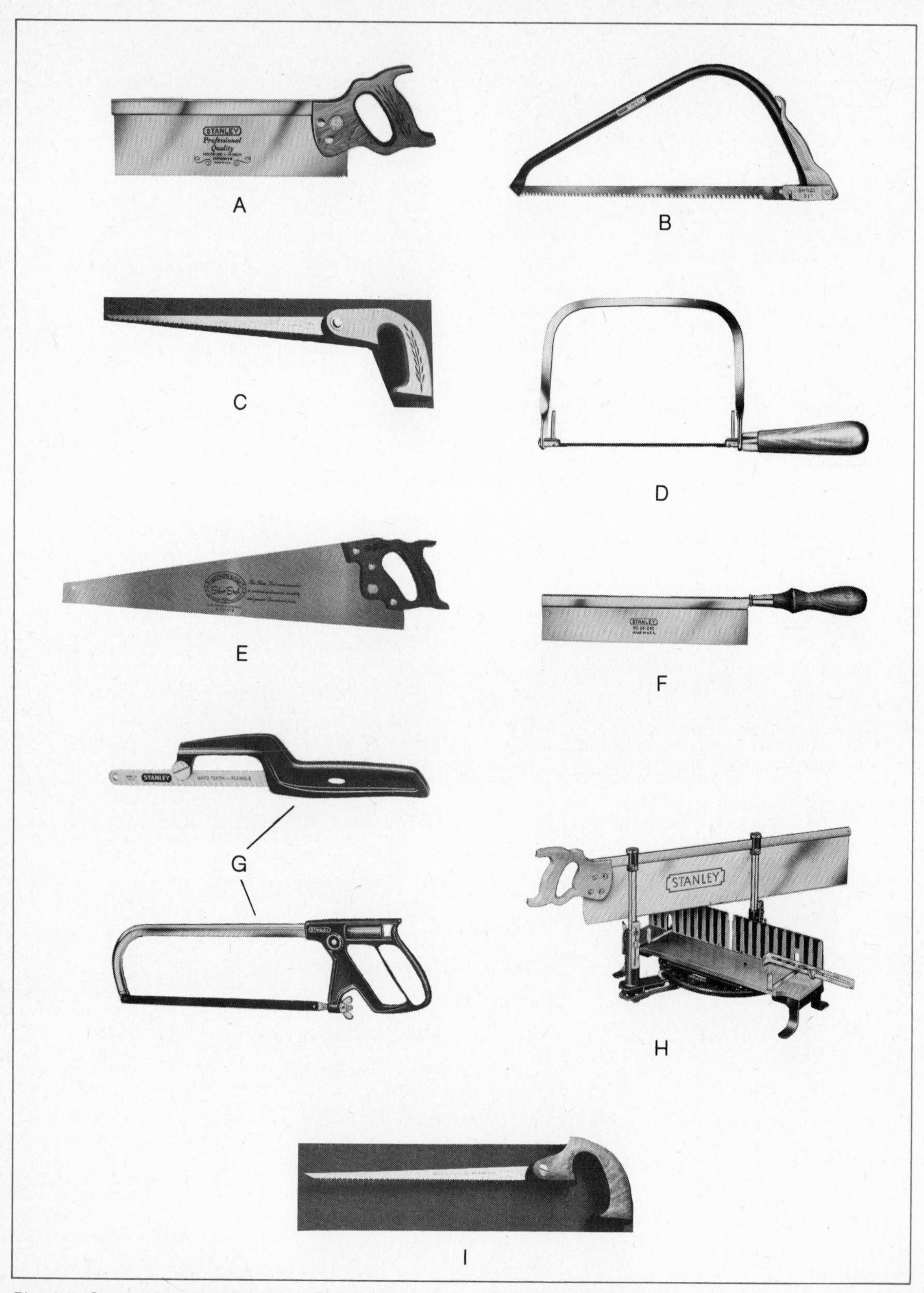

Fig. 6-1. Saws: (A) back; (B) bow; (C) compass; (D) coping; (E) crosscut and rip; (F) dovetail; (G) hacksaws; (H) miter saw and box; (I) keyhole.

6 Saws

6-1. General Description

Saws are used to cut a workpiece into one or more pieces, to cut an opening in a workpiece for cables, wire or pipes, or to make cuts such as tenons and mitered joints for joining workpieces together. Workpieces may be made of wood, metal, plastic or other materials. Different cuts may be made with different saw and blade combinations or with different techniques, such as using slow speeds.

Before describing a particular saw's characteristics, it is necessary first to define some terms common to all saws. The *butt* of a saw blade is the widest end, the end to which the handle is attached. The *tip* of the blade is the end opposite the handle. The *points* are the teeth, the coarseness of the blade being determined by the number of teeth – 5 points to the inch is considered coarse whereas 10 points to the inch is considered *fine.* Alternately set teeth are bent one to the left, the next to the right. The *kerf* is the cut made by the saw in a workpiece.

High-quality saws cut faster and more effortlessly because the teeth are precision-ground. A high number of teeth cut with a smooth finish; few teeth cut faster, but leave a rough finish on the workpiece. Saw blades are made of steel and stainless steel. Some are coated with Teflon-S* which is self-lubricating and highly resistant to abrasion. The coating helps the saw to cut smoother and also prevents rust.

Saw handles are made either of solid or laminated hardwood or of unbreakable, molded plastic.

The three most frequently used handsaws are probably the *crosscut,* the *hacksaw* and the *coping* saw. These saws should be among the first in your shop. Other saws you may consider having for your workshop and which are discussed in this chapter are: *back, bow, compass, dovetail, keyhole* and *rip.* The *miter box* that has wide application in cutting moldings is also discussed.

6-2. Application of Saws

Nearly every saw is used in a slightly different manner from other saws. Some cut on the push stroke, others on the pull stroke and still others on both strokes. Hence, general sawing techniques are discussed here; detailed descriptions of the use of particular saws are contained in the appropriate sections.

One of the most important considerations in any sawing operation is the support both of the workpiece and of the scrap being cut off. It is nearly impossible to make an accurate cut with any saw without adequate support. The type of support depends on the size of the workpiece. In cutting material from a piece of 4′ × 8′ plywood, for example, sawhorses or something similar, such as chairs or tables, are used. Smaller pieces may be held in special jigs or in wood vises. Metal pieces may be held in a machinist's vise that has its jaws protected to prevent marring of the workpiece surface.

After the workpiece is supported, draw a pencil mark completely across the workpiece where the saw cut is to be made. Place the saw to the scrap side of the mark at a 45° angle. Place the first knuckle of the thumb of your free hand against the blade edge. Using the butt part of the blade and your thumb as a guide, pull back on the saw and make a slight notch. Repeat until there is a small groove. Then take cuts the length of the blade. Let the saw cut at its own speed by allowing its weight to do the cutting. Don't push down on it – take your time. Keep the saw at an angle of about 45° for the most efficient cutting. (The ripsaw is an exception – it should be held at 60°.) Be sure to hold or support the scrap when it is ready to separate from the workpiece. Failure to do this will cause the end to split. Also, slow the speed and the length of the stroke near the end of the cutting.

If your job is to trim off only a slice of material from a workpiece, clamp a piece of scrap against the workpiece and cut both pieces off. If you plan to make a long, straight cut, clamp a piece of straight scrap along the marked cutting line and use the scrap as a guide for your blade. To cut at an angle, again clamp a piece of scrap wood, but clamp it parallel to and at the proper distance from the marked line to cut the required angle. In this application, tilt the blade against the scrap and start the saw into the wood at the mark. Continue to guide the saw along the marked line and against the scrap.

If the blade starts to swerve away from the cutting line, bend the blade slightly to bring it back onto line. Avoid sharp bends. If you go off the guide line quite a bit, return the saw to the point where you start to go off and saw straight. You can use a plane later to smooth off the miscut.

On long cuts, the saw may bind in the kerf. To alleviate this binding, wedge a piece of wood into the kerf. Move the wedge along the kerf, as required, to prevent binding.

To ensure that you are holding your saw square to the workpiece, use a try square against the workpiece and upward to the saw blade. Check occasionally during sawing and straighten the saw as required.

6-3. Care of Saws

After you use a saw, apply a light coat of oil to the blade before storing it. Keep the handle bolts tight at all times.

If a saw blade becomes clogged with sawdust or pitch, wipe the blade with a kerosene-soaked rag. If this doesn't remove the sawdust or pitch, then soak the blade in kerosene. After soaking it wipe the blade with a rag or, if necessary, scrape it.

Saw blade teeth can generally be sharpened four or five times with a file before the teeth need resetting. Generally, sharpening should be left to the professionals and you can locate one in the yellow pages of your telephone directory under SAWS – SHARPENING AND REPAIRING. Blades of the bow, compass, coping, hacksaw and keyhole saws should be replaced rather than resharpened. Specific procedures for sharpening crosscut and ripsaw blades are discussed in Sections 6-8 and 6-13.

6-4. Back Saw

The back saw is used in the workshop for cutting wood joints and for making straight cuts in molding. It is used most often as the saw in a miter box (Section 6-12). Because the back saw has a rigid, substantial back that prevents the blade from flexing, it is used to make straight cuts; it is never used for curved cuts.

The teeth of the back saw are finer than the teeth of either the crosscut or ripsaw. This gives it the ability to cut either across or with the grain of the wood. The blade is rather thick, with teeth that are bevel-filed and set; thus, when a cut is made, the teeth should cut just to the outside of the cutting line in the scrap of wood so that the workpiece is an accurate length.

Back saws are from 10″ to 30″ long; the short saws are used for general straight cuts where the workpiece is held in a wood vise or clamp. Saws used with miter boxes are from 22″ to 30″ long. The width of the blade under the back is from 2-1/2″ to 3-1/2″ for the short saws and from 4″ to 6″ for the long saws. There are 11 to 13 points to the inch.

The back saw should be used in a miter box (Section 6-12) if one is available and if the workpiece fits into the miter box. When a miter box is not used, clamp the workpiece so it is stationary. Start the cut with the saw at about a 45°angle and pull back until a starting cut is made. Then proceed back and forth, holding the blade at 45°. When you near the bottom of the cut, straighten the saw and cut straight.

6-5. Bow Saw

The bow saw is a lightweight tubular-frame saw used primarily to cut trees and logs, although it can also be used to make rough cuts on large pieces of wood such as posts, two-by-fours and planks. It is the ideal saw for cutting firewood in camp and is much superior to an axe. Blade sizes range from 21″ to 30″ and are easily replaced.

Support the piece of wood to be cut. In camp, place another log crosswise under the log to be cut and place your foot on it to clamp it tight. Keep your foot a safe distance from the blade. Hold the saw with one hand on the tension clamping lever and frame. Place the other hand on top of the tubular frame at a comfortable distance from the end. Place the blade at the cutting mark; draw the saw toward you and push it back only a few inches until the blade begins its cut. Then take long strokes – the length of the blade. Keep the blade flat and straight and prevent it from twisting. Let the saw do the cutting – no extra force is required. The teeth cut on both the push and the pull strokes and clear out the sawdust, leaving a clean kerf.

Keep a light coat of oil on the blade to prevent rusting caused by outdoor dampness. If a blade breaks, replace it as follows: pull the tension clamping lever out from the handle and around the pivot point 180° degrees. Remove the blade from the pin. On the other end, remove the sheet metal screw or bolt that holds the blade. Insert the new blade and reverse the above procedure.

6-6. Compass Saw

Craftsmen, electricians, plumbers and carpenters use the compass saw for making cutouts of curved and straight shapes in wood or plywood. It is often used to make openings in floors and walls for cables, pipes and electrical boxes. Its chief advantage is that it can start cutting from a small hole bored through the floor or wall. Its heavy-duty blade cuts wood and light metal.

The compass saw is from 12″ to 14″ long; there are from 8 to 10 points per inch in the blade. The blade tapers from about 1-1/2″ at the laminated hardwood handle to about 1/4″ at the tip. Some compass saws come with a nest of blades of different lengths and with different numbers of points per inch. Other models have a pistol grip with an index finger hole for a firm grip.

When using the compass saw, be sure to keep the blade cut just outside of the pattern line in the scrap area. This is necessary because the heavy-duty blade makes a rather wide cut. If necessary, as in removing material for an electrical outlet box, bore a hole of about 1/2″ diameter first. Using the saw tip, make vertical strokes to get the blade cutting along the pattern. Then tilt the saw to a 45° angle and continue at this angle until near completion of the cutout. Then return to the use of straight strokes.

Lightly oil the blades to prevent them from rusting. Remove and replace the blade by removing and reinstalling the screw-type fastener in the handle.

6-7. Coping Saw

The coping saw is used by carpenters, home craftsmen and hobbyists for cutting some otherwise impossible curves and angles in wood, plastic and thin metal. Curves of small diameters can be cut because of the thin, narrow blade. It is the only handsaw that is used for cutting intricate scrollwork. The coping saw is about 12″ long with a spring-steel, U-shaped frame that provides a throat opening from 4-1/2″ to 7″, depending on the model; a wide throat will permit cuts to the center of large workpieces. Blade lengths are 6-1/2″ with widths of 1/6″ to 1/8″. There are from 10 to 20 points per inch in the high-carbon steel replaceable blades. Because blades are inexpensive, they are replaced rather than sharpened when they become dull.

The angle of the blade is adjustable so that it can be rotated in the frame to allow the saw to do coping work over a larger work area. As can be seen in Figure 6-1, the blade is mounted between two pin-shaped handles that are perpendicular to the blade. To rotate the blade, loosen the hardwood handle by unscrewing it. When the blade is relatively loose, rotate the pin handle at the far end of the frame either clockwise or counterclockwise to the desired blade angle. Rotate the pin at the handle end. Hold this pin in position and tighten the handle until the blade is tight. On some models, the pin at the frame end is part of a notched stud that provides about 25 blade positions. These notches act as locks to hold the blade in position.

The blade is normally placed in the coping saw frame with the teeth pointing away from the handle. In this position, cutting is accomplished on the push stroke. However, there are times when you may desire to pull the saw toward you to cut; in this case, the blade must be turned around. The blade is replaced or turned around by unscrewing the handle until it is very loose. Remove the blade from the blade slots. Note that there is a tiny cross pin on each end of the blade that holds the blade into the saw frame. Place the new blade (or the reversed blade) into the slots. Rotate the pin handles to the desired angle and tighten the handle.

When using the coping saw, support the workpiece in a vise or clamp it to a work surface. Then place the blade to the cutting line and begin cutting. The coping saw blade cuts only in one direction (the blade cuts on the push stroke when the blade is inserted with the teeth pointing away from the handle). The beginning strokes should be short. Take full strokes after the blade has started along the pattern line on the workpiece. Keep the blade perpendicular to the work and follow the curved or straight lines of the pattern. If the frame comes to the workpiece, loosen and rotate the frame by way of the pin handles around the blade to the side of the workpiece to clear the frame. Thus, it is not necessary to remove the blade from the workpiece. Retighten the handle.

To make interior cuts in a workpiece, bore a 1/4″ hole through the workpiece. Remove the blade from the frame, place the blade through the hole in the workpiece and replace it in the frame. Tighten the handle and proceed with the cutting.

6-8. Crosscut Saw

The first handsaw that you should purchase is the crosscut saw. It is used primarily to cut across the grain of wood, but can also be used to cut with the grain, although with less efficiency. The crosscut saw is always used on plywood, regardless of the direction of the grain on the top or bottom ply. Crosscut saws are available in lengths from 16″ to 26″; the most popular sizes are 24″ and 26″. Widths of blades are from 6-1/2″ at the butt to 1-1/2″ at the tip. Blades are available with different numbers of points from 5 points per inch (coarse cut) to 10 points per inch (fine cut). The teeth are alternately set to the left and right; this allows the sawdust to be cleared from the cut and also makes the cut slightly wider than the thickness of the blade. The

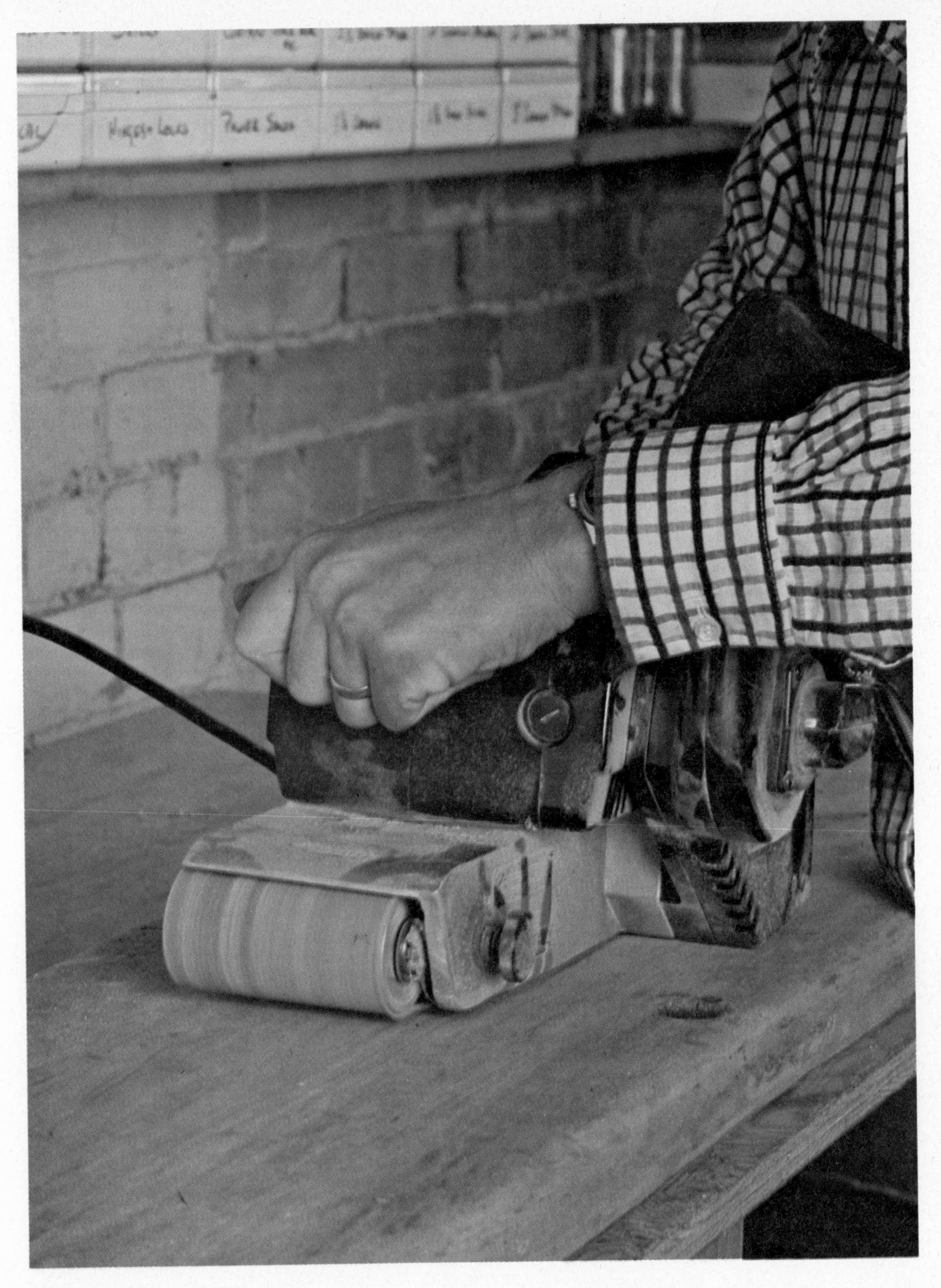

The belt sander is used for removing paint, stains and varnishes, for sanding rough lumber, and for smoothing badly scarred surfaces.

teeth cut on both the forward and backward strokes.

To cut a piece of wood, draw a line across the workpiece at the cut mark. Support the wood firmly for cutting. Place the saw blade on the scrap side of the cutting line at an angle of 45°. Using the butt part of the blade and the first knuckle of your thumb (on your free hand) as the blade guide, pull the saw and make a small starting notch. Repeat until a small groove appears. Then start to take longer cuts. Let the saw cut at its own speed by letting the weight of the saw cause the teeth to cut. Don't push on the saw – take your time. Keep the saw at a 45° angle for the most efficient sawing.

You should leave the sharpening of crosscut saws to a professional, but if you've got to try it, here's how: the crosscut points can be filed four or five times before they need to be reset. If resetting is required, however, it should be done prior to filing. Place the blade between two pieces of hardwood the length of the saw and clamp the wood and saw into a vise. The wood should be 1/8" below the gullets of the points. Stroke a flat file across the top of the points to flatten the teeth points. These flats act as guides in sharpening. Next, use a *saw set* to alternately bend each tooth to the left and the right. The alternate teeth on the one side are set first, followed by the teeth on the opposite side.

With the tip of the saw blade to your left, find the first tooth on the left end which is pointed toward you. Place a triangular file into the gullet to the left of this first tooth, position the file at 45°(Figure 6-2a), and file until one-half of the flat at the top of the tooth is gone. Continue alternately with every other tooth until you reach the handle. Turn the saw around in the vise and in the same manner file the other alternate teeth (Figure 6-2b) until the other half of the flat is removed.

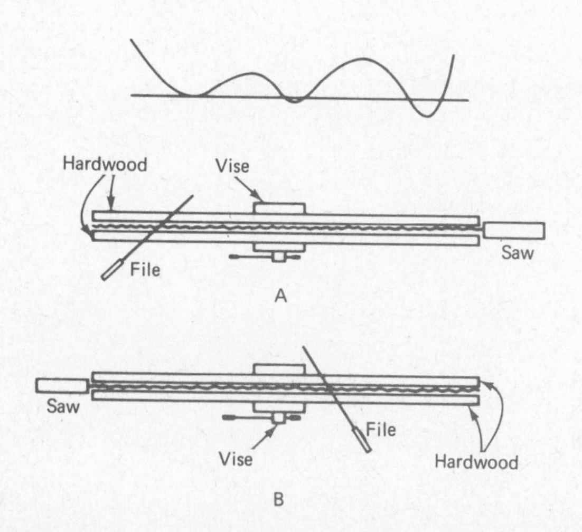

Fig. 6-2. Filing crosscut saw blade points.

6-9. Dovetail Saw

The dovetail saw is also known as a *hobby* or *cabinet* saw. It is similar in appearance, function and operation to the back saw (Section 6-4) except that it is smaller and the handle is round rather than a handgrip design. This smallness makes the dovetail saw a favorite with hobbyists and modellers for use in intricate dovetailing, tenoning, dadoing and rabbeting.

Like the back saw, the dovetail saw makes straight, true and smooth cuts. The blade leaves a narrow kerf. The dovetail saw is used in the same manner as the back saw although it is not used in a miter box.

The dovetail saw is about 10" long from the hardwood handle to the blade tip. The blade is 25-gauge polished steel and has a rigid back; the width under the back is about 1-1/2". Typical blades have 15 points to the inch.

6-10. Hacksaw

A hacksaw, consisting of a frame, a grip and a blade, is used to cut metal sheets, iron pipe, tubing, screws and bolts, steel bars, brass, bronze, copper and other metals. The frame puts tension on the blade. Most frames adjust to accommodate both 10" and 12" blades, have notched studs to hold the blade in four positions and unbreakable plastic grips. The throat depth is from 3" to 4". A wing nut and tension wheel adjust the blade tension. The blade is mounted with the teeth forward to cut on the forward stroke. A mini-hacksaw that uses regular hacksaw blades or broken blades is used when the regular hacksaw is unsuitable. The mini-hacksaw is 9″ long (without blade).

Hacksaw blades are available in 14, 18, 24 and 32 teeth per inch. The 18 teeth per inch blade is recommended for general, all-purpose work. The 14 teeth per inch blade is used for most metals with sections of one inch or more thickness; 18 teeth per inch for steel bars, brass and copper 1/4″ to 1″; 24 teeth per inch for iron pipe and medium tubing 1/8″ to 1/4″ thick; and 32 teeth per inch for thin metal sheets and tubing up to 1/8″. In general, then, fine teeth are used for thin stock and coarse teeth for thick stock. Remember the following rule: you have chosen the proper blade if two of its teeth can rest simultaneously on the workpiece (or on the wall of a tube). If two teeth cannot be put on the material, tilt the hacksaw. Various blade widths are available for cutting different-sized slots, as for slots in bolt heads. There are also two teeth sets: *wavy* for thin work and *regular* for general, all-purpose work.

There are high-speed and standard blades. Blades classed as high-speed, shatterproof and flexible provide a balance between toughness and flexibility which assures safety and cutting efficiency. Blades made of molybdenum are hard, long-wearing and straight-cutting and are preferred by toolmakers for cutting rigidly clamped material. The molybdenum blade is also ideal for general machine shop use. Tungsten blades cut the toughest materials. Tungsten blades are also used on tool and die steel and abrasive materials.

Standard flexible blades for general cutting have only the teeth hardened; thus, the back remains flexible to minimize breakage. This type of blade is economical for the inexperienced or occasional user. Flexible blades are also used in cramped areas without a vise.

Special unbreakable standard blades, used for miscellaneous cutting, are balanced, safe, tough and flexible. The blade is first hardened throughout and then made flexible, a process that results in hard, sharp teeth with a semi-hard back to withstand the twisting that causes blade breakage and stripping of the teeth.

The standard all-hard blade is for the skilled technician. Used for workpieces which can be held in a vise, it gives exceptional uniformity and cutting ability. It is important that the workpiece being cut be held rigid to prevent the blade from breaking.

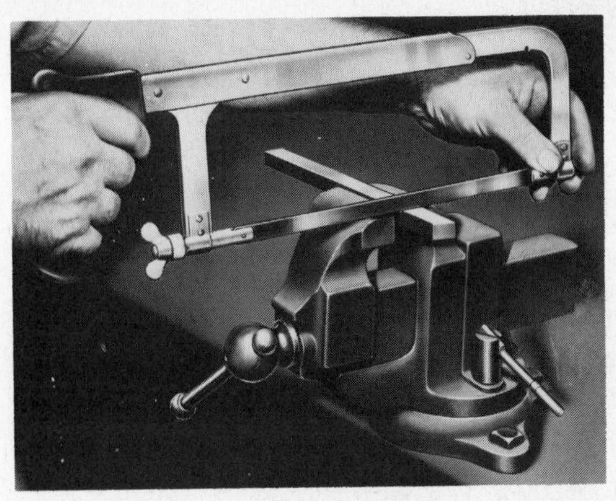

Fig. 6-3. Proper use of the hacksaw.

Two special blades are available for the hacksaw. A tungsten, carbide-tipped blade is used in cutting ceramic, steel and glass; it is not used for soft metals. A spiral-toothed blade enables the workman to use the hacksaw like a coping saw.

Place the workpiece firmly in a vise or other holding jig. Use a file to make a small starting nick on the workpiece. Then hold the saw firmly with one hand on the grip and place the thumb of the other hand at the guideline. Place the blade on the guideline and against the thumb. Use light pressure and slow, short strokes as the hacksaw cuts on the guideline. Take strokes of 1″ to 2″; apply light, downward pressure on the forward stroke. Do not apply pressure on the back strokes: the blade does not cut on this stroke and any pressure in this direction only dulls the blade. After the cut has been started, remove your thumb and place it on the saw (Figure 6-3). Increase the stroke length to 6″ or 7″. Do not twist the blade as it may break. When the cut is nearly through the workpiece, use light pressure and slow, even strokes until the workpiece parts. If the workpiece is larger than the throat of the frame, loosen the tension wheel and wing nut and rotate the blade to the side or bottom to continue the cut. For overhead cuts, many users invert the blade. Lubricants or coolants are generally not used. When cutting pipe or

1 *Using the table saw and the electric drill to make a rabbet joint with dowels.*

2

3

4

rod, you should wedge the cut open if the blade starts to bind. When cutting thin sheet metal, sandwich the metal between two pieces of wood and cut through the entire sandwich. To cut thin-walled tubing, insert a dowel into the tubing.

Use the proper blade for the workpiece being cut. Whenever possible, grip the workpiece in a vise; when this is not possible, use a flexible blade. Keep the blade from becoming excessively hot during use. Replace dull blades.

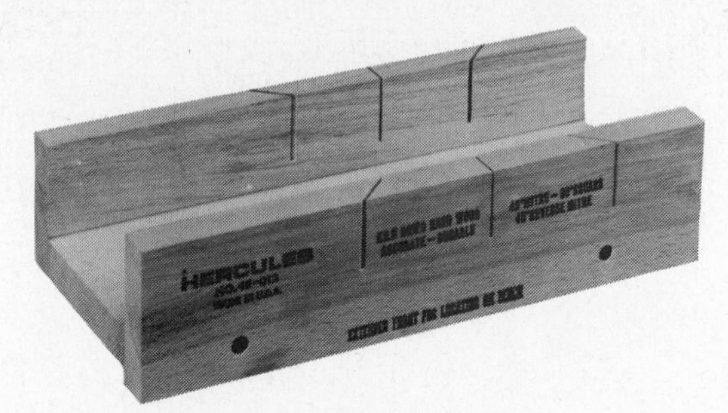

Fig. 6-4. The wooden miter box is used by the home owner to make accurate miter cuts in molding and picture frame woods.

6-11. Keyhole Saw

The keyhole saw is similar to the compass saw (Section 6-6) except that it is smaller and its blade is narrower; it can therefore cut smaller diameters than the compass saw. The keyhole saw cuts holes for electrical cables, water and gas pipes, and fixtures in walls and floors. Keyhole saws are from 10″ to 12″ long and have 10 points per inch. The keyhole saw is used in the same manner as the compass saw.

6-12. Miter Box

The miter is an abutting surface or bevel on either of the pieces joined in a miter joint. A miter joint is formed when two pieces of identical cross section are jointed at the ends, and where the joined ends are beveled at equal angles. A miter box is used to cut these miters. Miter boxes can be made of economical hardwood or they can be made of metal with many features to assure accuracy and cost over a hundred dollars. Obviously, your needs determine the type of miter box you'll choose. For home use, one of hardwood is adequate; a craftsman making handcrafted picture frames and other moldings may choose the more sophisticated model to ensure accuracy.

The wooden miter box (Figure 6-4) is made of 3/4″ kiln-dried oak, rock maple or other hardwood. A typical box is from 11″ to 16″ long and has a cutting depth and width of 1-3/4″ and 3-1/2″, respectively. It has saw blade guide slots cut into it at 45° and 90°.

The back saw (Section 6-4) is used to cut wood placed in the miter box. Place the miter box on your bench with the lip extended over and fixed tight against the edge (Figure 6-5). Place a piece of scrap wood on the bottom of the box to protect it and to raise the workpiece toward the top of the box. Place the workpiece on top of the scrap and against the back of the box. Align the marked cutting line with the slots so that the blade will cut just outside of the mark. If possible, use a clamp and another piece of scrap wood to clamp the workpiece tight against the back of the box. Otherwise, hold the workpiece tight against the back with your hand. Place the saw into the appropriate miter box slots and against the workpiece with the handle tilted slightly upward. Start with a pull stroke. As the saw goes slightly deeper, level the saw blade until it is parallel to the workbench. Take long strokes and let the weight of the saw do the cutting. Go slowly and easily at the completion of the cut.

Figure 6-1 illustrates a miter box made of malleable iron and a back saw used by professionals. The positive saw guide controls the saw blade and also holds the blade up so that the workpiece can be positioned in the miter with both hands. The rear blade guide pivots about an index plate that is graduated in degrees. The plate is also marked for 4-, 5-, 6-, 8-, or 20-sided figures. A spring-loaded pointer stops at the most frequently used angles.

Adjustable stock guides are used to hold

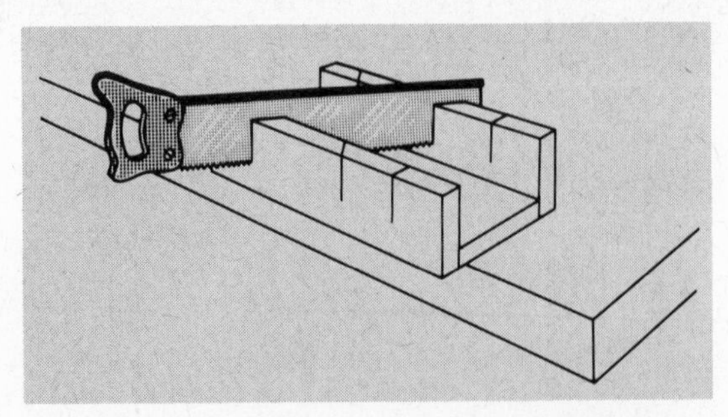

Fig. 6-5. The miter box is used to cut mitered joints in moldings and other decorative pieces.

the wood against the back of the miter box. An adjustable length-stop is used to set up a standard workpiece length for duplicate pieces. Stops are also provided for sawing to a given depth and to prevent sawing below the baseboard.

The capacity of width of the workpiece is related to the angle of the cut. For example, the maximum width of a workpiece that can be cut with a 26″ × 4″ back saw at 90° is 8-5/8″; at 45°, 6″; and at 30°, 4″.

More practical all-metal miter boxes for the home craftsman are available. A sliding key secures the saw guide at the proper angle as well as at the desired depth of cut. The saw guide may be set at the most commonly used angles of 90°, 60°, 45° and 30°.

To use the metal miter box, raise the back saw to the top. Set the box to the desired cutting angle. Place the workpiece against the back and clamp it with the stock guides. Adjust the length-stop guide if you're planning to cut more than one workpiece. Lower the back saw carefully to the workpiece. Begin the cut with a pulling stroke and continue with long strokes until the workpiece is parted.

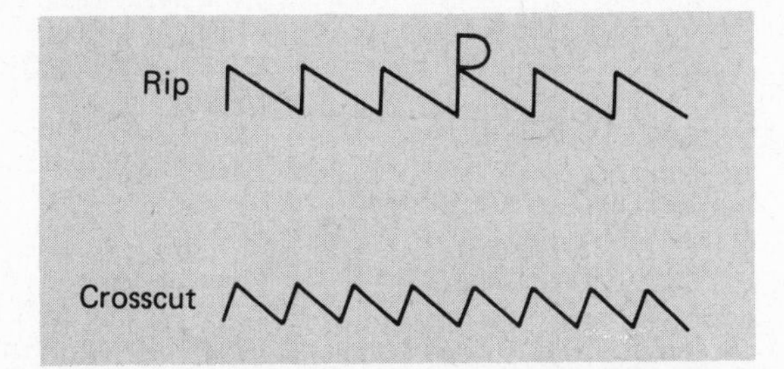

Fig. 6-6. Rip and crosscut blade teeth patterns.

All parts of high-quality metal miter boxes can be replaced with new parts.

6-13. Ripsaw

The ripsaw is the handsaw used to cut (rip) with the grain of the wood. The teeth of the ripsaw are shaped like tiny chisels with their edges crosswise to the saw. The teeth are alternately set to the left and right to clean out the sawdust. The ripsaw cuts only on the push stroke and operates most efficiently when it is held with the handle upward at an angle of 60°. A typical ripsaw is 26″ long and has 5-1/2 points to the inch.

You can identify an unknown saw blade as either a rip or a crosscut blade by examining the teeth and remembering the following: the ripsaw teeth are shaped the same as the angle formed at the bottom of a capital letter R standing for Rip (Figure 6-6).

In cutting with the ripsaw, support both the workpiece and the scrap piece. Hold the saw at the cutting line with the handle upward at an angle of 60°. Place the blade tip on the workpiece against the first knuckle of the thumb of your other hand. Make sure that the blade is going to cut just to the outside of the line marked on the scrap wood. Make a few short strokes to get the saw started (remember that the teeth cut only on the push stroke), then continue with strokes that are the length of the saw. It may be necessary to place a wedge of scrap wood into the saw kerf to prevent the workpiece from binding the saw. For long ripping cuts, you can clamp a length of straight scrap wood along the cutting line. Saw along the edge of the scrap wood, using it as a guide.

The sharpening of ripsaw blades should be left to professionals, but if you want to try it, here's how: proceed in exactly the same manner as with the crosscut blade (Section 6-8), but hold the file straight across the blade. File every other tooth; then turn the saw around and file every tooth not previously filed. The procedure of turning the saw around prevents you from removing more metal from one side of the blade than from the other.

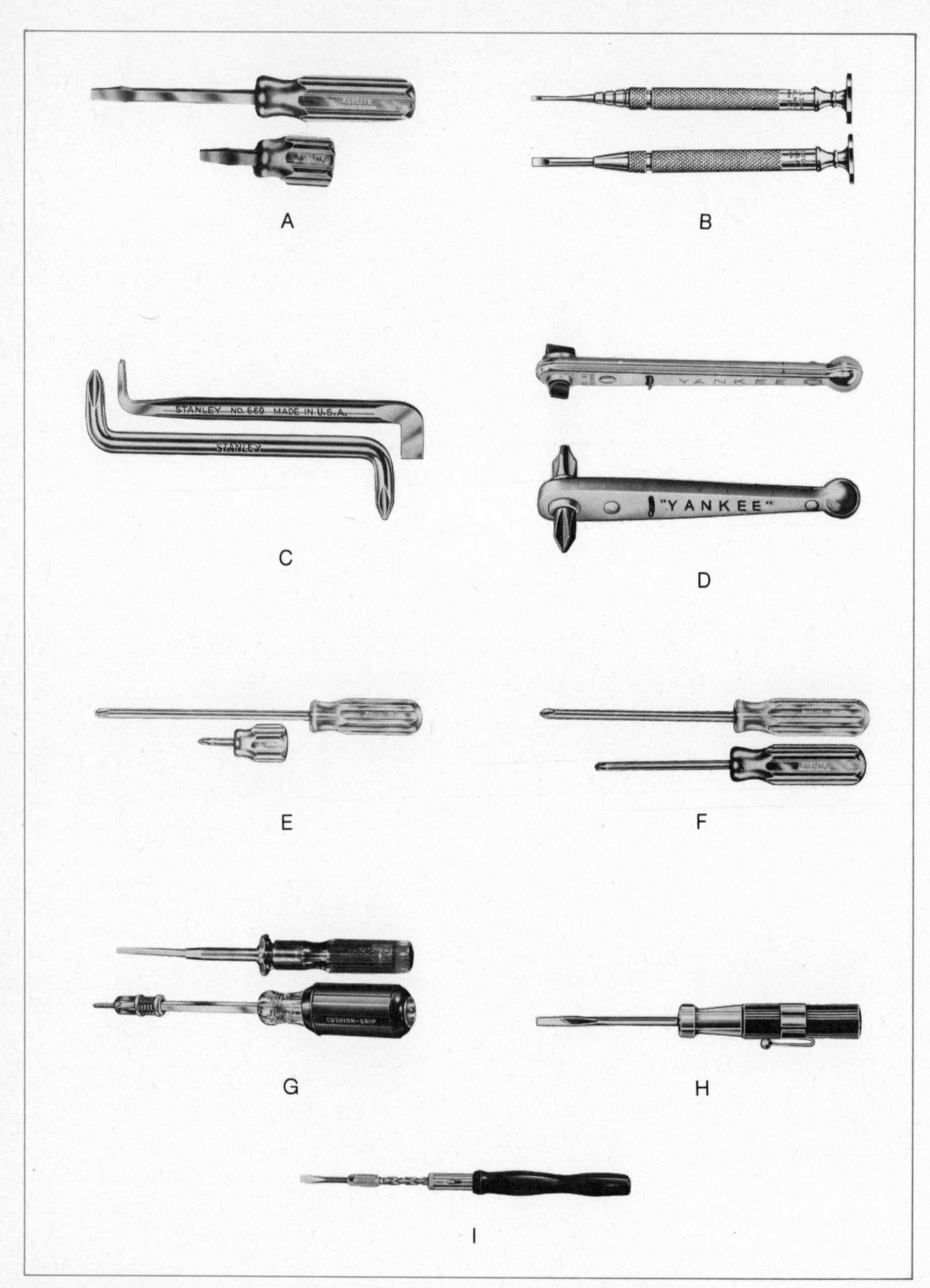

Fig. 7-1. Screwdrivers: (A) conventional; (B) jewellers'; (C) offset; (D) offset ratchet; (E) Phillips head; (F) Reed and Prince; (G) screw-holding; (H) spark-detecting; (I) spiral ratchet.

7 Screwdrivers

7-1. General

Screwdrivers are named either for the type of tip they have – *conventional* (for slotted screws), *Phillips* or *Reed and Prince* – or for their shape or function – *offset, jewellers', offset ratchet, screw-starting, spark-detecting* or *spiral ratchet.*

Screwdriver blades are made of tempered steel alloy and are available with chrome plating to retard rusting. The tip end is cross-ground and the sides are hollow-ground. The blade is anchored deep in a shockproof, breakproof plastic handle. Wide flutes on the handle provide a firm gripping surface.

Screwdriver bits for use in braces are also available. This type of screwdriver is particularly useful for installing a large number of screws or for installing screws where a large amount of torque is required.

7-2. Application of Screwdrivers

Although the screwdriver is a simple tool, it is often misused: be careful to select the longest-bladed screwdriver which is convenient and which has the proper tip for the job. A long screwdriver allows for added pressure and control on the screwhead. The width of the conventional tip should be the same as that of the slot of the screw or bolt; the thickness of the tip should equal the width of the screw or bolt slot (Figure 7-3). Undersize screwdriver tips will cause damage to the fastener, to the tip itself, and possibly, by slipping, to the workpiece. Oversize tips will likewise mar the workpiece when the screw or bolt head comes flush with the workpiece. The Phillips and Reed and Prince tips should fill the fastener slots.

In use, the screwdriver is gripped firmly by the handle; the blade is steadied and guided in the screw or bolt slot with the other hand. Always keep the blade in the center of the screw or bolt head and apply turning pressure straight through the blade. Tiny screwdrivers may be held between the thumb and index finger (Figure 7-2).

Use screwdrivers only for their intended purpose. Never use them for prying or chiselling. Never use pliers on the blade of the screwdriver; if additional torque is required, use a square-bladed screwdriver with a proper-sized wrench to fit the blade. Never hammer the end of a good screwdriver because you may damage the tip. If you need to remove paint or rust from a screw slot, use an old screwdriver and angle it in the slot. Lightly tap the handle.

To install small screws, make screw-start-

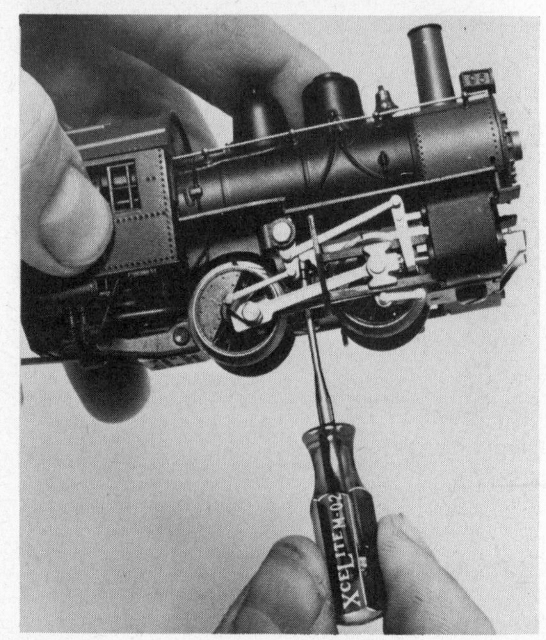

Fig. 7-2. Tiny screwdrivers used in equipment repair or model work are held with two or three fingers.

ing holes with an awl. For large screws, use a hand drill or electric drill. Screw threads may be lubricated with soap to aid installation.

7-3. Care of Screwdrivers

If the tip of a conventional screwdriver becomes worn or nicked so that the tip no longer fits the screw or bolt slot, the tip must be reground. When using a grinder, adjust the tool rest to hold the screwdriver against the wheel until the desired shape is produced. Use an emery wheel. Square the tip, then grind both sides of the tip until it is the required thickness. Keep the blade thick enough to make a fairly tight fit in a screw slot. The sides should be nearly parallel. During grinding, frequently dip the screwdriver into water to prevent overheating and consequent loss in the steel temper. (If the tip turns blue during grinding, the metal has lost its tempering and it will be necessary to grind beyond this point for maximum strength.)

If the tip of a Phillips or a Reed and Prince screwdriver becomes burred, place the blade in a vise and carefully remove the burr with an oilstone. If the tip is sufficiently worn so that it does not properly fit the screw or bolt head slots, then the screwdriver must be replaced.

7-4. Conventional Screwdriver

The conventional screwdriver is probably the tool most often used around the house. It is suggested that you buy one good set to use exclusively for tightening and loosening screws and bolts and also some 8″ to 10″ inexpensive, conventional screwdrivers for opening cans, prying, tightening eyebolts and chipping. Keep an inexpensive spare in your glove compartment as a general-purpose tool for use in emergency automobile repairs.

The conventional screwdriver has a blade from 4″ to 12″ long; short screwdrivers are called *stubby* screwdrivers and are used in tight spaces. Screwdriver blades may be round or square; a wrench may be used on a square blade to enable you to apply more torque to the screw or bolt. Conventional screwdrivers are sold in sets with blade tip widths, blade lengths, and handles proportioned for the most frequently used screws and bolts.

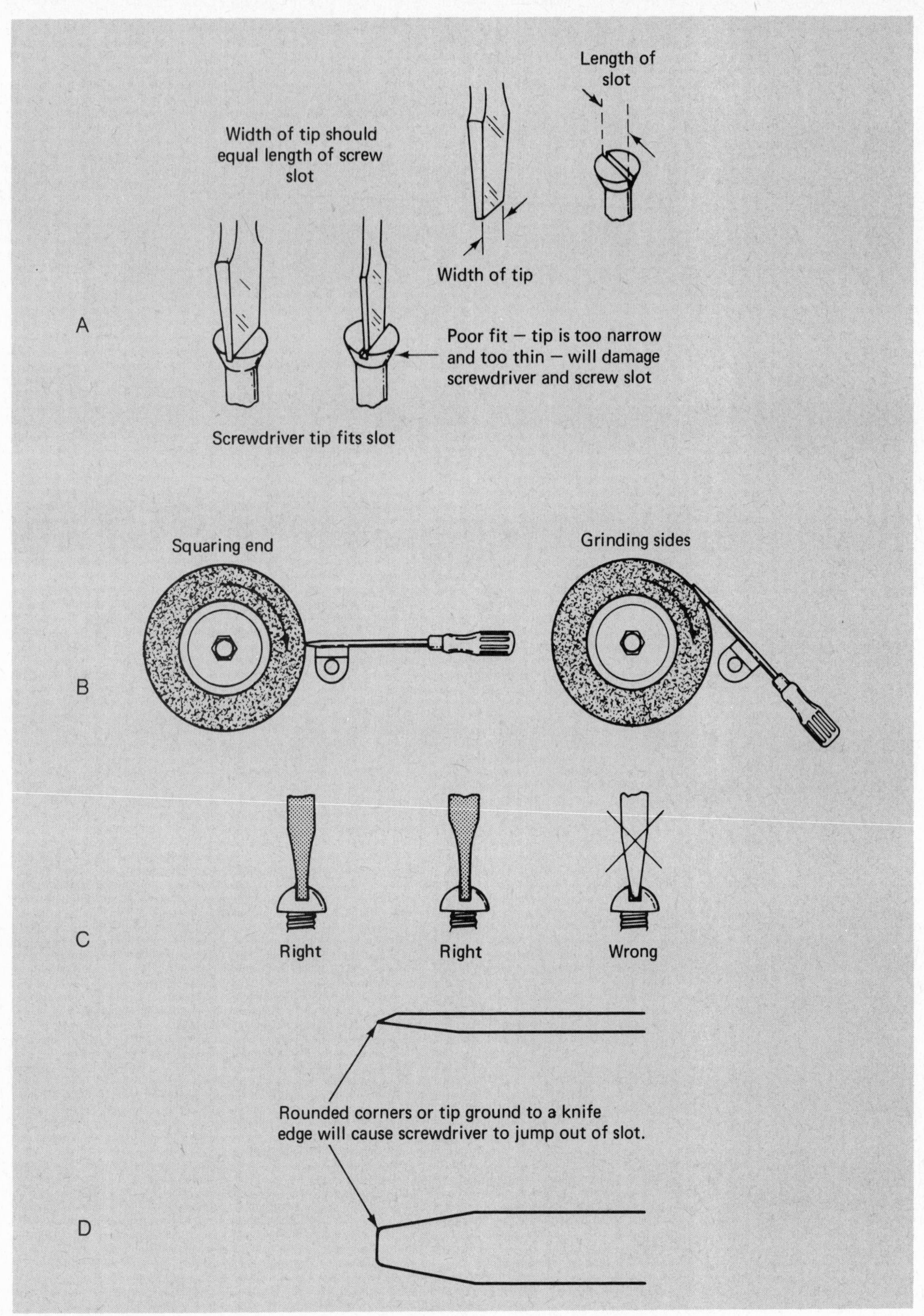

Fig. 7-3. Grinding conventional screwdriver blade tip.

The drill press stand has many attachments, one of which is shown on the facing page.

Drum sander attachment on drill press.

7-5. Jewellers' Screwdrivers

Jewellers' screwdrivers are for delicate, fine work. They are especially suitable for the model railroader, for example, or for parents who have occasion to repair small toys. Precision jewellers' screwdrivers are used by watchmakers and clock makers, jewellers, opticians and toolmakers. Jewellers' screwdrivers of various sizes are available to the professional in sets, but the occasional user should buy a set that consists of one body with four or more interchangeable screwdriver blade bits. A typical set includes four blades with conventional tips from 0.025″ to 0.050″ plus one Phillips head blade tip. Each blade is notched or keyed so that it fits into the body and locks, preventing rotation during use; a knurled chuck tightens the blade into the body.

Select the desired blade and insert it into the chuck. Rotate the blade until you feel the blade notch slip into place. Tighten the chuck. Place the blade into the screwhead, press your index finger on the concave swivel knob and rotate the body to tighten the screw (Figure 7-4).

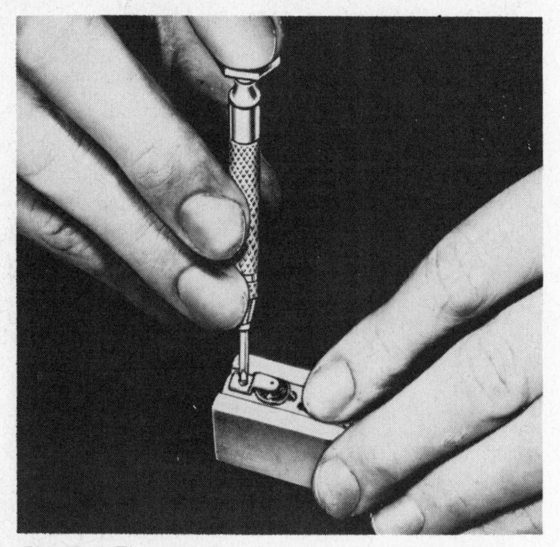

Fig. 7-4. The jewellers' screwdriver is used for screws and bolts on models, small toys, or for precision work such as watch repairing and toolmaking.

7-6. Offset Screwdriver

The offset screwdriver is used to drive or withdraw screws or bolts that cannot be aligned with the axis of the conventional or Phillips head screwdrivers because of space limitations, such as when mounting an electrical box to a wall. The offset screwdriver is also used where high torque is required to tighten or loosen a screw or bolt. Offset screwdrivers are made in a variety of sizes with different-sized conventional and Phillips heads. On models with four tips, two are offset at 90° and two at 45°. This offset allows for quarter- or eighth-revolution torquing.

7-7. Offset Ratchet Screwdriver

Like the offset screwdriver, the offset ratchet screwdriver is used where space limitations prevent the use of the straight-drive screwdriver and where high torque is required. The advantage of the ratchet, of course, is that the user may tighten a screw without removing the screwdriver blade tip from the screw. A lever allows ratcheting in either direction for tightening or loosening. Models are available either with two different tips in conventional or Phillips head or with combination conventional and Phillips heads. Also available are sets which have an offset ratchet handle with various sizes of slots, Phillips head, hollow-head setscrew, cap screw and bolt adapters.

7-8. Phillips Head Screwdriver

A Phillips head screwdriver is distinguished by its tip, which is designed to fit into Phillips head screws or bolts. The tip has about 30 flutes and a blunt end. The screwdriver tips are made in four sizes to match Phillips head screws as follows: size 1 for no. 4 and smaller screws, size 2 for nos. 5 to 9, size 3 for nos. 10 to 16 and size 4 for nos. 18 and larger. Blade lengths vary from 1″ to 18″.

The tip of the Phillips head screwdriver is similar to the tip of the Reed and Prince screwdriver (Section 7-9). The Phillips screw has beveled walls between the slots, whereas the Reed and Prince screw has straight, pointed walls. The Phillips screw slots are not as deep as the Reed and Prince. Be sure you are using the proper screwdriver for a given screw by holding the screw in a vertical position with the head up, and placing the screwdriver tip in the head. If the screwdriver tends to stand up unassisted, it is probably the right type and size.

7-9. Reed and Prince Screwdriver

The Reed and Prince screwdriver is similar to the Phillips screwdriver except that the tip is defined as a cross-point. The tip has 45° flutes and a sharp, pointed end. The outline of the end of the Reed and Prince screwdriver is close to a right angle. This type of screwdriver is used on *Frearson* screws. Blade lengths vary from 3″ to 12″.

7-10. Screw-Holding Screwdriver

Screwdrivers are used to hold and start slotted screws into a workpiece. This allows you to start driving screws with one hand or in confined spaces. Various screw-holding arrangements including steel jaws and wedgelike split-tip screwdrivers are available. The steel jaws model has a set of clamped jaws that slide down the bit and grasp the screw (or bolt) under the head. The clamp holding the jaws is spring-loaded and thus holds the screwhead slot against the blade tip. The split-tip screwdriver wedges the blade firmly into the screw slot when the sliding sleeve is in position. Screw-holding screwdrivers are made in several sizes and must be matched to the size of the screw to be driven.

7-11. Spark-Detecting Screwdriver

The spark-detecting screwdriver is useful to the home auto mechanic. A tough composition handle contains a neon tube which glows when the screwdriver tip is touched to a high-frequency circuit such as those in auto ignition systems. Blade lengths are from 5″ to 9″ with tips from 9/64″ to 1/4″ wide.

7-12. Spiral Ratchet Screwdriver

To save time and labor, the spiral ratchet screwdriver may be used when a large number of screws must be driven by hand. A spring holds the interchangeable blade in the screwhead and automatically returns the handle to the driving position after each stroke. The screwdriver is operated by pushing on the handle; the spiral ratchet drives the screw.

By means of a lever on the screwdriver the user selects either right-hand spiral ratchet drive for tightening, left-hand spiral ratchet drive for loosening or a rigid setting which allows him to lock the spiral drive mechanism. The latter setting makes the tool into an ordinary screwdriver for final torquing of a screw or bolt.

Spiral ratchet screwdrivers are usually sold with two screwdriver bits: a conventional bit and a Phillips head bit. Various-sized models are sold for special applications: light-duty for small parts assembly, all-purpose regular duty (7/32″ or 1/4″ conventional bit and a no. 2 Phillips head bit), and heavy-duty.

Spindles and chucks, shell and handle tubes, shifters, springs and small-piece parts of the spiral ratchet screwdriver can be purchased and replaced without special tools or procedures.

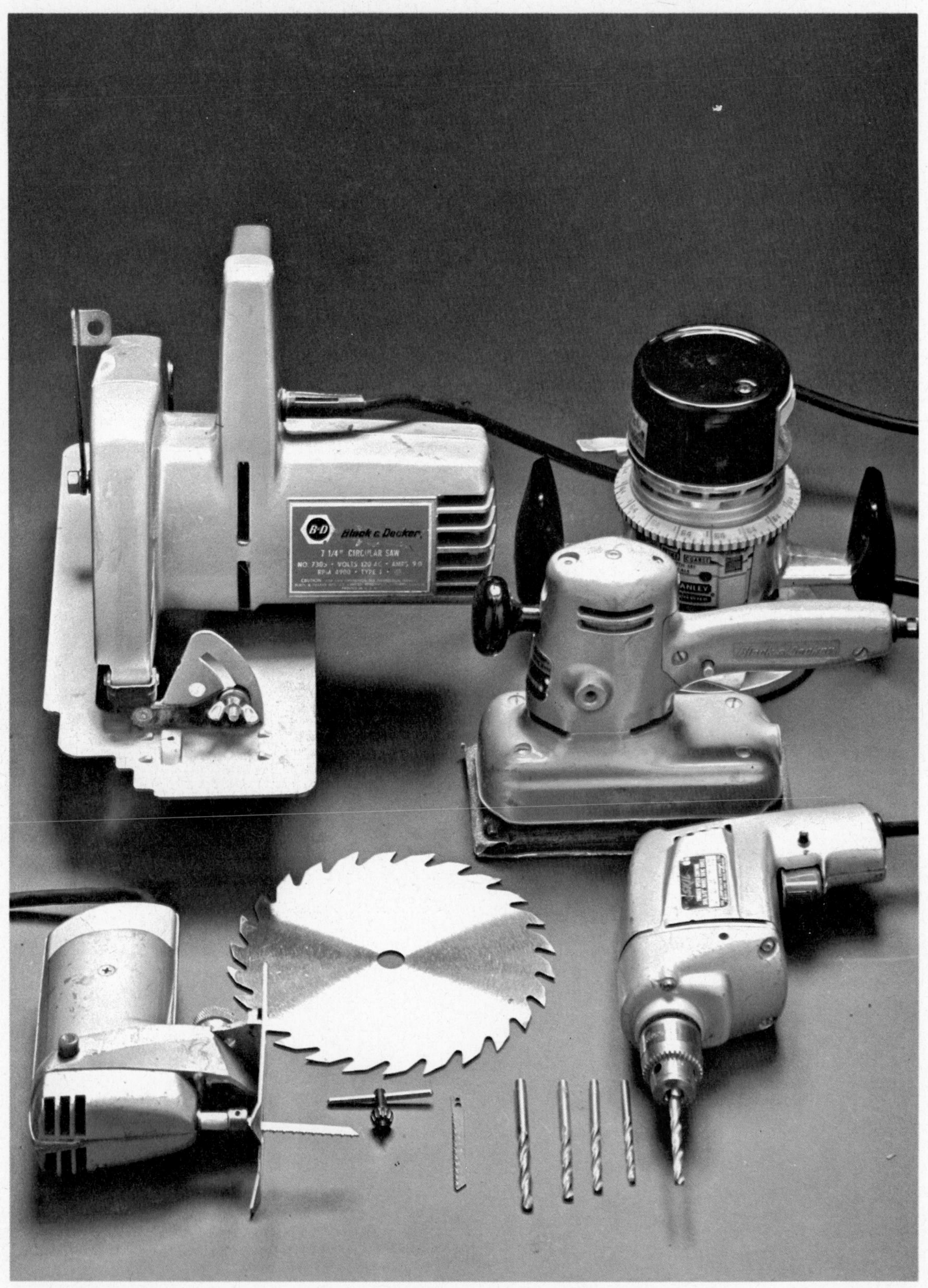

Clockwise from top left: circular saw, router, orbital sander, electric drill with bits, and jigsaw (or sabre saw).

Wrenches

8-1. General Description

One variety of wrench is used to tighten or loosen fasteners such as bolts, nuts, headed screws and pipe plugs, while another is made to grip round stock, studs and pipe. There are also special wrenches such as locking plier, torque and setscrew wrenches: the locking plier wrench is used for a multitude of jobs; a torque wrench is used to tighten square or hex-headed fasteners to a specific tension; and setscrew wrenches are used for tightening special hex setscrews.

There are many types and sizes of wrenches for various applications. You should select and purchase those wrenches most applicable to your needs. For example, if you're a new home owner, you might select an adjustable open-end wrench and a locking plier wrench as your first tools. They can do a multitude of jobs for you. If you're a beginning mechanic, you'd probably select a socket wrench set and a set of combination wrenches. An electronic technician would certainly want a set of nut-drivers among his tools. And the automobile do-it-yourselfer would want an adjustable open-end wrench, an oil filter wrench and a socket wrench set in his toolbox. The hobbyist would select miniature wrenches. Purchase only high-quality wrenches for your toolbox or shop. Cheap ones slip, bend and deteriorate rapidly.

Wrenches are made of hardened and tempered steel. Many are chrome-plated to prevent rusting. On fixed, nominal opening wrenches (such as open-end, box, combination and sockets), the size of the opening is stamped near the opening; sizes are available in English units and metric units. The openings are from 0.005″ to 0.015″ larger than the size marked on the wrench, so the wrench fits easily over the nut or bolt.

This chapter describes the following wrenches: *adjustable open-end, box, chain pipe, combination, locking plier, nut-driver, oil filter, open-end, pipe, setscrew, socket, strap* and *torque.* Determine your needs, purchase the wrenches and learn to use them properly.

8-2. Application of Wrenches

In selecting a wrench for a particular job, consider the type of work to be done, the limitations of work space and the number of fasteners to be used. There are not any hard and fast rules as to which wrench should be used for a particular job, but there are some general guidelines. When a wrench cannot be used over a fastener head, an open-end wrench of either the fixed or adjustable type

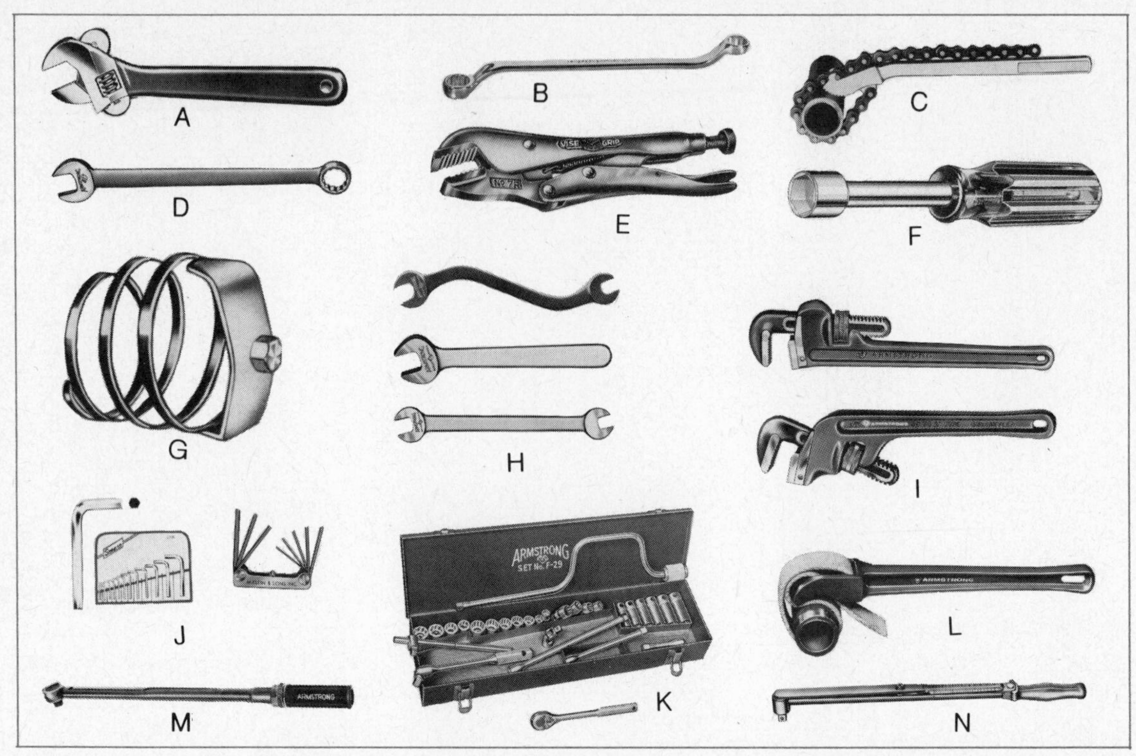

Fig. 8-1. Wrenches: (A) adjustable open-end; (B) box; (C) chain pipe; (D) combination; (E) locking plier wrench; (F) nut-driver; (G) oil filter; (H) open-end; (I) pipe; (J) setscrew; (K) socket; (L) strap; (M) torque-limiting type torque wrench; (N) deflecting beam/ converging scale type torque wrench.

should be used. Thus, the open-end is used on nuts of fuel, oil and hydraulic lines, clutch and transmission control rods, brake rods and cable ends. A box wrench is used where the wrench *can* be placed over the fastener head. Many advanced technicians prefer combination wrenches – they use the open end to spin the nut down and the box end for the final torquing.

Select the proper-sized wrench for the nut, bolt or other fastener to be loosened or tightened. If you use a wrench that is the wrong size, the points of the nut or bolt will become rounded, making it very difficult to tighten or remove. Always use adjustable wrenches in such a way that the pulling force is applied to the fixed jaw and not to the adjustable jaw. Never use a pipe or other device to extend the length of the handle of a wrench; instead, obtain a wrench with a longer handle. Do not strike wrench handles with a hammer to tighten or loosen a bolt or nut.

If a rusted nut, bolt or other type of fastener is to be removed, apply penetrating oil or a mixture of kerosene and lubricating oil and allow time for the oil to penetrate before attempting to turn the fastener with a wrench. Rather than push, pull on the wrench whenever possible to avoid skinning your knuckles if the wrench slips. Do not exert a hard pull on a pipe wrench until it has a firm grip on the workpiece.

8-3. Care of Wrenches

To keep wrenches from being damaged, always use the proper-sized wrench. Ensure that adjustable open-end, pipe, chain pipe, strap, oil filter and locking plier wrenches are adjusted to fit properly; this protects the wrench and the workpiece.

Always keep wrenches free of oil, grease, grit and dirt. After using them, clean them with a rag saturated with kerosene. Avoid getting

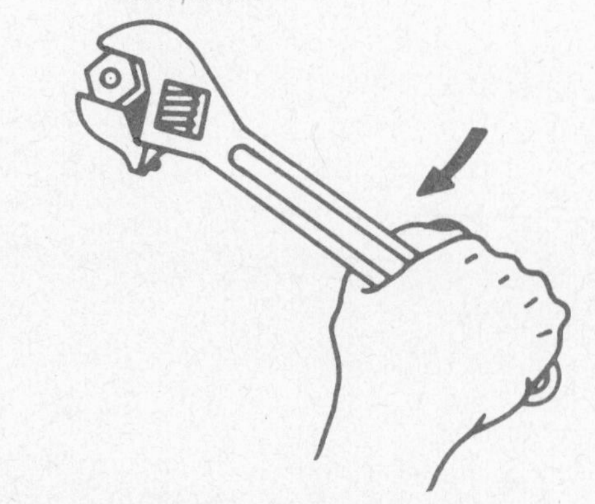

Fig. 8-2. Fitting and using the adjustable open-end wrench.

grit into the working parts of ratchets, chains and adjustment nuts. Remove grit at once; a few particles can score your tools deeply.

The knurled nuts of adjustable open-end and pipe wrenches should be lubricated occasionally with graphite. To prolong the life of your wrenches and assure their better operation, apply a drop of light oil to the areas of moving parts when required.

8-4. Adjustable Open-End Wrench

An adjustable open-end wrench is used to tighten an infinite number of English-size and metric-size nuts, bolts, headed screws and pipe plugs. It is normally used when an open-end, box or combination wrench has one fixed and one movable jaw and is shaped so that the jaws form four sides for firm gripping of hex nuts or bolts. The wrench head is thin and angled from the handle for use in tight areas. Since its jaws are smooth, it is widely employed for tightening and loosening chrome-plated fittings. The angled head has proved so useful that this wrench has taken over the functions of the monkey wrench to such a degree that there are few monkey wrenches available today. The adjustable open-end wrench is also known as the *Crescent wrench,* after its first manufacturer.

The movable jaw of this wrench is adjusted by rotating a knurled nut. Some models include a locking device in the form of a push rod to hold the adjustment. Typical wrench handles are from 4″ to 16″ long with jaw capacities of approximately 1/2″ to 1-3/4″. Long handles provide the best leverage.

Always place the adjustable open-end wrench on the nut or bolt to be tightened or loosened so that the force used is applied to the fixed jaw as shown in Figure 8-2. After placing the wrench in position, tighten the knurled adjustment nut until the jaws fit the fastener tightly. If the jaws do not fit tightly, the wrench will slip and the corners of the nut or bolt will become rounded (your knuckles may also be skinned). Press the locking device (if available on your wrench) to hold the opening. In tight areas, the wrench can be flipped over repeatedly to give you better leverage on the fastener.

8-5. Box Wrench

Box wrenches completely surround the nut or bolt and hence are safer than the open-end wrench. The box opening contains 6 or 12 points; the 12-point end allows the wrench to be used for 1/12 turns (30°) in close quarters. The box wrench is usually a double-ended wrench with different-sized openings in each end. The openings range from 1/4″ to 2-3/4″ and 8 to 32 millimeters. Midgets from 3/16″ to 11/32″ are available. The height of the box is correctly proportioned for all nut sizes and series. Handle lengths vary from 4-1/2″ to 18″; the long lengths provide greater leverage. The handle may be straight, but it is often offset 15° from the box opening to allow for hand clearance. Sometimes the box end is offset. Ratchet box wrenches are also available, but are recommended only for mass production operations. Always select the proper-size box wrench (Figure 8-3) for the nut or bolt to be turned.

Repeated swings in arcs of 30°(a 12-point

wrench) are sufficient to completely tighten or remove a nut or bolt. Remove the wrench completely from the nut or bolt and then reposition it for another partial turn. This method may be slow, but it enables you both to make tight fittings and to loosen tight fittings. Thus, it is often desirable to *spin down* a nut or bolt rapidly with an open-end wrench and then set it tight with a box wrench. For this reason, many home owners and mechanics prefer the combination wrench (Section 8-7).

8-6. Chain Pipe Wrench

The chain pipe wrench is a unique wrench in that it can be used on any shape stock – square, hexagonal, irregular or round. (Figure 8-1 shows the chain pipe wrench encircling a round workpiece.) It can also be used in close quarters, such as around a pipe that is against a wall or under a sink and can't be gripped with a pipe wrench (Section 8-12). It is used to tighten, loosen or grip various sizes of stock, pipe, nuts and plugs from 1/2″ to 6″ depending upon chain length. It can also be used to remove automobile oil filters. However, do not use this wrench on surfaces which you do not want marred: the serrated gripper can cause damage. Home owners, mechanics, plumbers, electricians, steamfitters and maintenance men will find many uses for this tool.

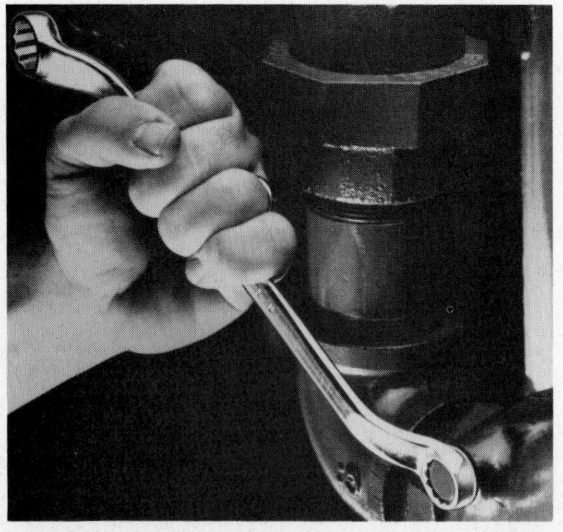

Fig. 8-3. Always use the proper size box wrench.

To use the chain pipe wrench, place the curved surface that is just below the serrated gripper tightly against the workpiece. Wrap the *sprocket chain* around the workpiece as tightly as possible and lock the *chain sprocket* in the slot behind the serrated gripper. Pull the wrench so that the workpiece is forced between the serrated gripper and the chain. Apply a leverage force to the wrench handle so that the gripper bites into the workpiece and the wrench turns it. Some models have a double set of serrated gripper edges which allow use of the wrench in either direction without removing it from the workpiece.

Keep the chain well lubricated with oil during storage to prevent rust.

8-7. Combination Wrench

The combination wrench has an open-end wrench at one end of the handle and a 12-point box wrench at the other; nominal openings at both ends are the same. The open end is often angled at 15°from the handle and the box end is often offset 15° from the plane of the handle to give hand clearance. Combination sets are made in various sizes of openings, lengths, offsets and angles as discussed for the open-end and box wrenches. The combination wrench is often the choice of home owners, mechanics and technicians because it combines the open end for quick spin down of nuts or bolts and the box end for final tightening (see Sections 8-5 and 8-11).

8-8. Locking Plier Wrench

A locking plier wrench provides a vise-like

grip around any nut, bolt, flat or round object placed between its serrated jaws. It is actually seven tools in one: clamp, gripping tool, pipe or locking wrench, pliers, adjustable wrench, portable vise and a wire and bolt cutter. In essence, it's a pinch hitter where other tools fail. The upper jaw is fixed; the lower jaw is accurately adjusted to the desired jaw opening by turning the knurled adjustment screw. A locking handle fastens the jaws to an object with up to a ton of pressure by the toggle action of a lever being pushed past center. The lock release lever opens the jaws to allow the removal of the wrench. Typical wrench jaws open up to 2″; the locking capacity is from 1-1/4″ to 1-3/4″ for wrench lengths from 7-1/2″ to 10″. The 5″ locking plier wrench is a handy tool for small jobs.

This wrench can serve as a handle for files, chisels, screwdrivers, hot pans or faucets. It can be used to manipulate sheets of steel, plastic or wood, or to improvise tools such as an awl by simply clamping a heavy needle or nail in its jaws. A scraper can be made with a razor blade clamped between the jaws. It is an excellent clamp for use in gluing, soldering, riveting, welding, sawing, filing or cutting any material. A saw may be fashioned by clamping a broken blade or a small blade into the jaws. It can be used as a plier to bend, form, twist, hold and crimp metal and wire. It's also a small, portable vise and can be locked into a larger vise when needed to hold small objects for polishing, filing, grinding, soldering. It will pull out headless nails, remove broken studs, and even become an adjustable open-end and pipe wrench. In summary, the locking plier wrench is a necessity for

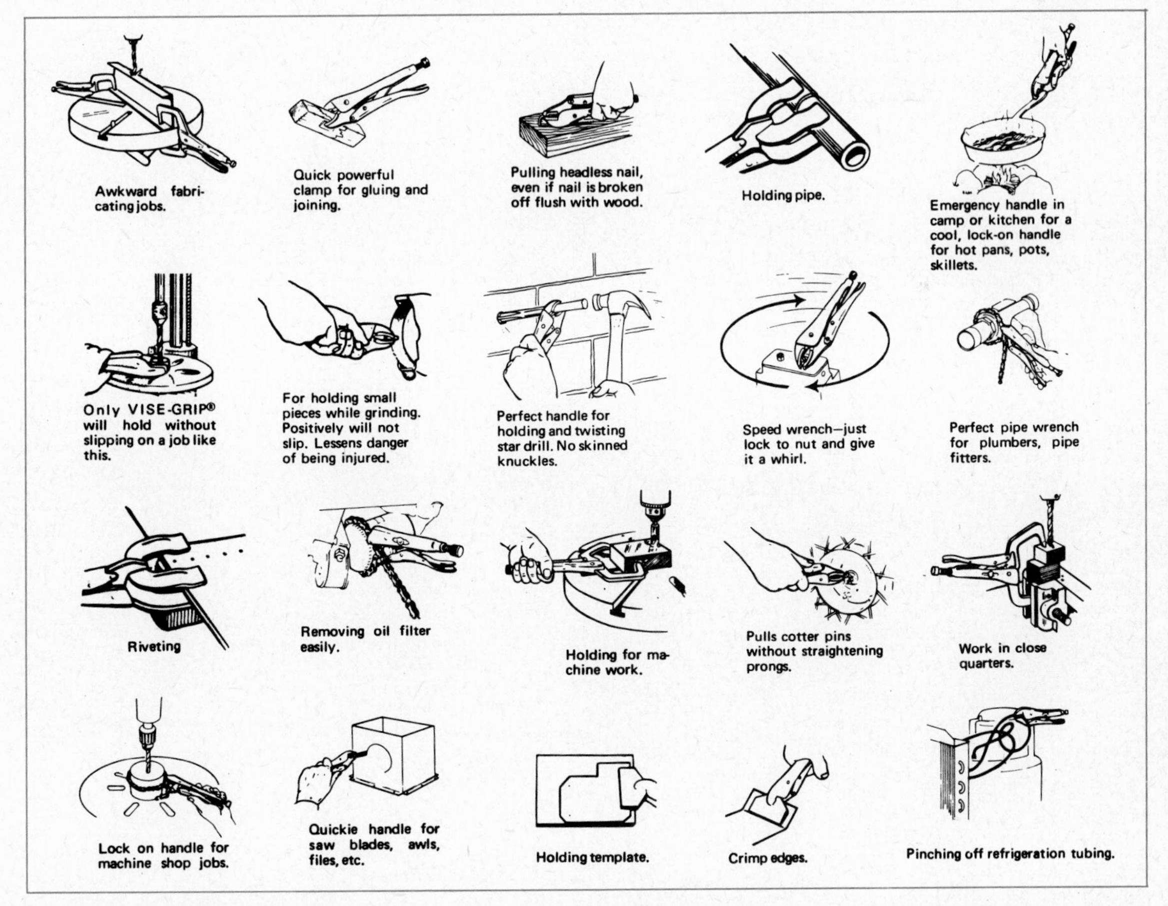

Fig. 8-4. The locking plier wrench and other similarly designed tools are used for numerous applications in the home owner's shop as well as for professional applications.

numerous applications and should be in every toolbox or shop (Figure 8-4). It's also handy to carry in your automobile for emergency repairs.

Several other tools valuable to the home owner, technician, machinist and welder that are similar in design and operation to the locking plier wrench are illustrated in Figure 8-5. The advantage of the curved-jaw locking plier wrench over the straight-jaw (Figure 8-1) is that it is superior for use on pipes and other round objects. The curved jaw contacts hex nuts and bolts at four points for extra gripping power. The wire cutter (located below the jaws) can cut a quarter-inch bolt in two when successive bites are taken.

The C-clamp is used for clamping angle iron or I-beams, holding work onto a drill press table, holding workpieces for gluing, welding and soldering, and for numerous other clamping purposes. Unlike a screw C-clamp, this C-clamp is adjusted to the correct opening before use. It is then tightened by a squeeze of the hand for a locking grip. In factory tool work, it holds templates and parts for die work.

The sheet metal tool is designed primarily for use in sheet metal work for clamping, bending and fabrication. Furniture repairmen and upholsterers find this tool very versatile and valuable.

The chain clamp uses the same locking and leverage principles as the other tools, but its jaws are a chain and a body hook. An emergency tool for immobilizing odd-shaped pieces, it will hold any part or parts, regardless of their shape, so long as its chain will wrap around them.

The welding clamp allows the user to align the edges of pieces to be welded. Then it locks the pieces together and thus frees the user's hands. The U-shaped jaws increase the user's visibility and working space for the welding operation.

To grip an object with any of the locking plier wrenches or similar tools, grasp the tool with the locking lever slightly open. With the other hand, rotate the adjustment screw until the jaws just slip over the workpiece. Squeeze the handles together until locked. After the job is completed, hold the wrench with one hand and hook the fingers of the other hand around the screw end of the wrench. Push the lock release with the thumb. To use as pliers without locking, adjust the end screw so that the jaws don't quite snap shut on the workpiece. Practise using the locking plier wrench on workpieces of various sizes until you are fully acquainted with the procedures for locking and unlocking it. The surfaces of some objects may be marred by the jaws and clamping pressure of these tools. Protect the surfaces with scraps of wood, soft metal or other material.

Damaged serrated jaws or adjustment screws can be renewed by filing. Place the jaw or screw into a vise and carefully file with a three-square file. If the spring housed in the fixed handle becomes stretched or broken, it may be easily replaced by following the manufacturer's instructions. Occasionally place a drop of oil on all points of wear.

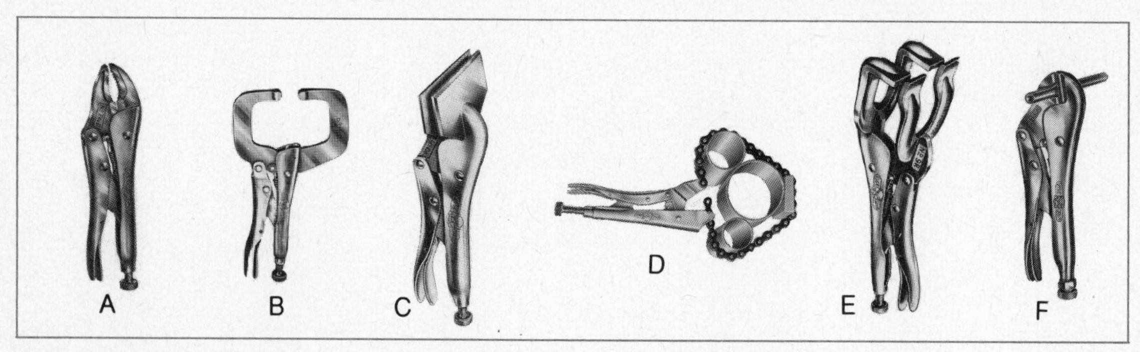

Fig. 8-5. Several other types of tools similar to the locking plier wrench are: (A) curved-jaw locking plier wrench; (B) C-clamp; (C) sheet metal tool; (D) chain clamp; (E) welding clamp; (F) pinch-off pliers.

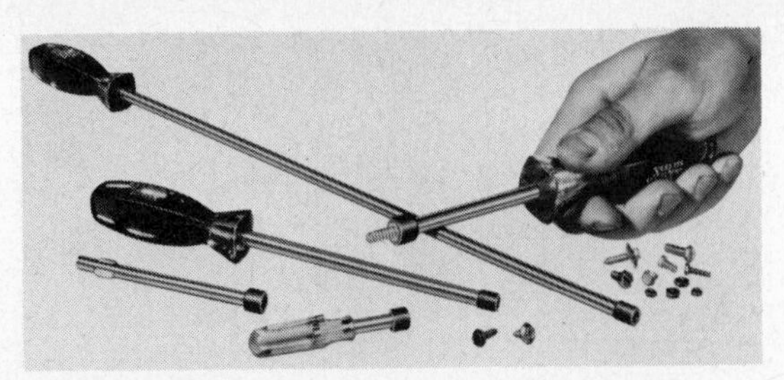

Fig. 8-6. The nut-driver wrench is used like a screwdriver.

8-9. Nut-Driver Wrench

The nut-driver is similar in appearance and use to a screwdriver except that the tip is a six-point socket for hexagonal nuts and bolts. The nut-driver speeds up the turning of nuts. The lower part of the shank is often hollow to accept the end of a bolt as the nut is turned down onto the bolt. There are nut-drivers to fit sizes from 3/16″ to 5/8″. The handles are shockproof butyrate plastic or wood and are fluted for sure gripping; the handles are often color-coded according to size. Nut-drivers are available in midget pocket clip, stubby, regular and extra-long styles.

The nut-driver is used a great deal by the electronic technician to spin nuts and hex-headed sheet metal screws rapidly. It is used like a screwdriver (Figure 8-6). After the nut is tightened, be sure that the nut-driver does not round the corners of the nut and likewise, that the corners of the nut-driver do not become stripped.

Variable-size nut-drivers hold 1/4″ to 7/16″ hex nuts and bolts for easy starting and removal when you are working in tight spaces. The positive locking device holds the fastener. The harder you turn, the tighter it grips. You should buy this tool rather than a set of fixed nut-drivers if you have only an occasional need for it.

8-10. Oil Filter Wrench

Oil filter wrenches are used by the home automobile repairman and the garage mechanic to remove standard *spin-on* oil filters. Force applied to the wrench hex nut is distributed over the surface of the oil filter to prevent the splitting or crushing of the filters. The continued force of the wrench grips the filter securely to loosen and remove it. This wrench is not used to replace the filter.

Place the oil filter wrench over the oil filter; attach a hex socket, open-end or box wrench to the hex nut and apply turning power to remove the filter.

8-11. Open-End Wrench

The open-end wrench is usually a double-ended wrench with different-sized openings at each end. Wrenches are available with openings from 1/4″ to 1-3/4″ and 6 to 28 mm. Midget wrenches have openings ranging from 13/64″ to 3/8″. The common open-end wrench has ends which are angled at 15° to permit complete rotation of 30° hex nuts.

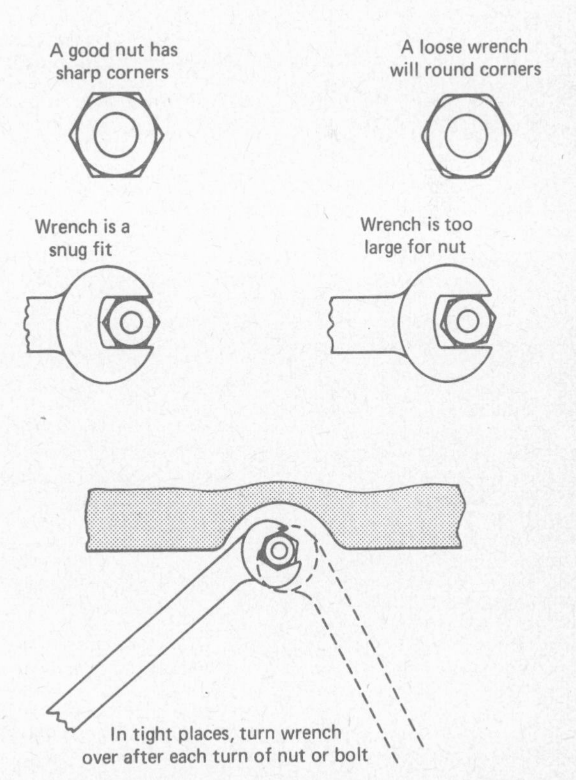

Fig. 8-7. Fitting and using the open-end wrench.

There are other open-end wrenches having ends at 45°, 60°, 75° or 90°. The length of the wrench is proportional to the size of the opening; the greater the length, the greater the leverage. Other types of open-end wrenches are single open-end or have S-shaped handles.

It is important to be sure that the open-end wrench fits the nut or bolt tightly so that the wrench does not slip and round off corners (Figure 8-7). Offset open-end wrenches make it possible to turn a bolt or nut that is recessed or in a tight place. The wrench may be turned over, often after each partial turn, to aid in turning the nut or bolt in limited areas. If possible, always pull the wrench toward you to move the nut or bolt in the desired direction. (If you must push, push with your palm so your knuckles won't be skinned if the wrench slips.) To avoid slipping during the final tightening of the nut or bolt, make the final turns with a box wrench.

8-12. Pipe Wrench

A pipe wrench (often called a *Stillson* wrench) is used to rotate a round workpiece such as a pipe or the head of a worn nut or bolt that has had its corners rounded; thus, the pipe wrench is not to be used on good nuts and bolts. The pipe wrench has two nonparallel, hardened jaws with milled teeth. The outer jaw, which is adjustable by means of a knurled nut, is called the *hook jaw.* This jaw is made with a small amount of play which provides an automatically tight grip on a pipe when the wrench is turned in the direction of the movable jaw. The marks made by the jaws can be reduced by placing a rag against the jaws. The healjaw, which is the fixed jaw, is replaceable. It is also spring-loaded to give it full floating action to increase the wrench grip. Pipe wrenches used by home repairmen, plumbers and mechanical technicians are available in straight and offset patterns in lengths from 6″ to 14″ with capacities of 1/4″ to 6″. The 10″ to 14″ sizes are best for most applications around the home.

The pipe wrench will work only when turned in the direction of the opening of the jaws. Force is applied to the back of the handle. Since the top jaw has slight angular movement, the grip on the workpiece is increased by the pressure on the handle. Correct direction of rotation for the pipe wrench is the same as for the adjustable open-end shown in Figure 8-2. In plumbing work, pipe wrenches are used in pairs: one holds the threaded pipe and one turns the fitting.

To replace the healjaw, cut the rivet with a cold chisel. Replace the jaw and rivet it into place.

8-13. Setscrew Wrench

Setscrew wrenches, also called *Allen wrenches,* and hex keys are L-shaped. They are made of tool steel and have a hexagonal section. The setscrew wrench is used to tighten or loosen socket setscrews, socket-head capscrews, button head capscrews, socket-head shoulder screws, flat head capscrews and "tru-round" pressure plugs having hexagonal sockets. These fasteners are used to hold handles, secure knobs and controls, and to lock pulleys to shafts. Dimensions across the flats are available from approximately 0.028″ to 2″ in English and metric units. A typical set includes wrenches of (in inches): 0.028, 0.035, 0.050, 1/16, 5/64, 3/32, 7/64, 1/8, 9/64, 5/32, 3/16, 7/32, 1/4, and 5/16. In addition to L-shape, setscrew wrenches are available with loop handles, as a flexible driver, and in a holder similar to a pocketknife.

When using the setscrew wrench, be sure that the proper-sized wrench is selected and inserted as far as possible into the fastener to be tightened or loosened. The long end of the wrench is used to turn a fastener rapidly, while the short end is used either for final tightening or for unfreezing a tight fastener.

8-14. Socket Wrench (With All Drives)

A socket wrench consists of a drive handle and one of various English or metric sockets.

The socket wrench is made from parts contained in a socket wrench set: sets may contain as few as six pieces or as many as 220 pieces, including drives, sockets, universal joints, adapters and other related tools.

A basic socket wrench set usually consists of 12-point sockets of various sizes, a ratchet drive, a flex handle, a slide bar handle and an extension. The socket wrench drives are available in 1/4″, 3/8″, 1/2″, 3/4″ and 1″ sizes and should be selected according to their required use. A 1/4″ drive is for light-duty work; 3/8″ and 1/2″ for automotive, industrial and aircraft work; and 3/4″ and 1″ for heavy-duty use including machinery and heavy equipment installations. A 3/4″ socket wrench ordinarily refers to a 3/4″ detachable socket plus a drive handle which is usually either a ratchet drive handle or a flex drive handle.

There are four basic types of socket wrench drives: ratchet, flex handle, speeder and slide bar handle. These drives attach to a variety of sizes of sockets which may be 4-point or 8-point sockets for square head nuts and bolts and 6-point or 12-point sockets for hexagonal head nuts and bolts. Accessories such as socket adapters, universal joints and extensions extend the capability of socket wrenches.

The ratchet drive (Figure 8-8a) is used with various-sized English and metric sockets to tighten or loosen nuts and bolts. The ratchet drive saves time because the socket does not have to be removed from the nut or bolt. It is particularly useful in close quarters; a fine-tooth action ratchet drive permits handle movements in increments of 4°.

Ratchet drives contain two and sometimes three controls. The quick-release push button releases spring pressure on the spring-tensioned ball in the square drive. The socket slides easily off the drive; a different socket readily slips on and locks over the spring-tensioned ball. The reversing lever allows the ratchet to slip in one direction for tightening and in the opposite direction for loosening. A knurled speeder on the ratchet allows nuts to be quickly spun onto a bolt. The speeder accessory may be added to a ratchet drive if it is not on the original model (a speeder is not shown in Figure 8-8).

The flex handle drive (Figure 8-8b) is used with various-sized English and metric sockets to tighten or loosen nuts and bolts. With the flex handle drive perpendicular to the work, nuts or bolts are first made snug. The flex handle is then turned 90° to provide additional leverage for final tightening of the fastener. In confined areas where the flex handle cannot

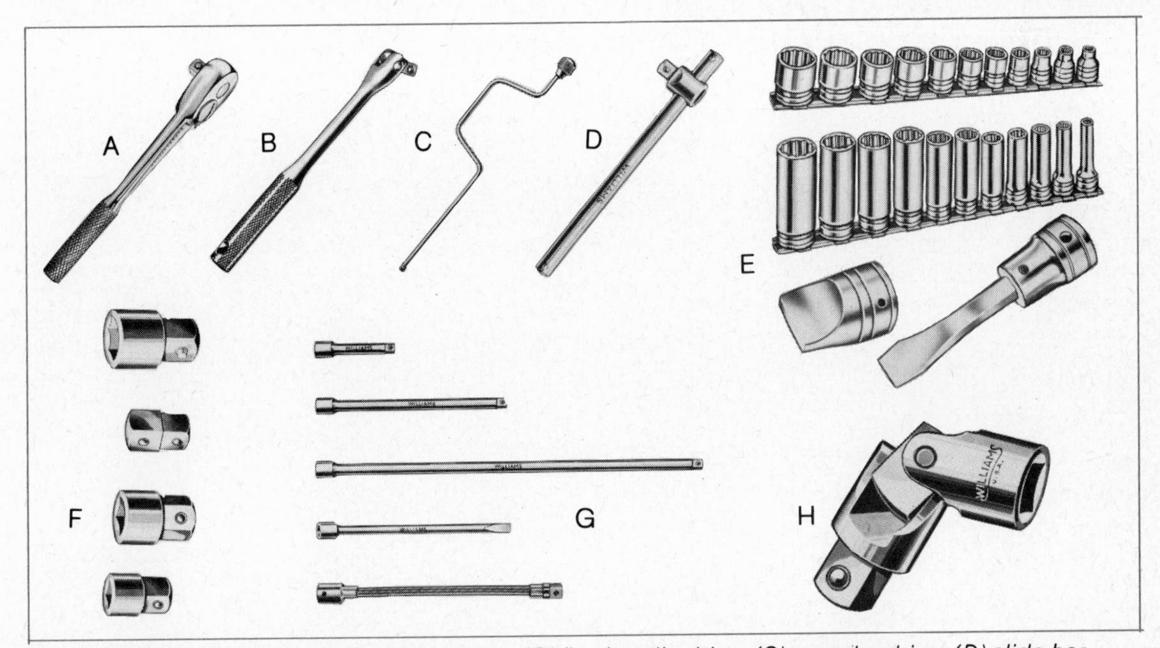

Fig. 8-8. Socket wrench set: (A) ratchet drive; (B) flex handle drive; (C) speeder drive; (D) slide bar handle; (E) sockets; (F) socket adapters; (G) extensions; (H) universal joint.

be turned 90°, a sliding bar may be placed through a cross hole in the handle: this too provides additional leverage for final tightening.

The speeder drive (Figure 8-8c) is used with various-sized English and metric sockets as a rapid means of spinning (running) a number of nuts and bolts on or off. Final tightening or breaking of fasteners must be done with the ratchet or flex handle drives that provide additional leverage.

The slide bar handle, often called a T-handle or L-handle drive (Figure 8-8d) is used with various English and metric sockets to tighten or loosen nuts and bolts. For additional leverage and for working in confined areas, the drive may be positioned anywhere along the handle; thus, a T-shape or an L-shape is formed.

Sockets (Figure 8-8e) have two openings: a square hole that fits the socket wrench drive and a circular hole with notched sides to fit the nut or bolt head to be turned. The square hole is 1/4″, 3/8″, 1/2″, 3/4″, or 1″ across to mate with the respective socket wrench drive. The circular hole may be notched with 4 or 8 notches for square nuts and bolt heads or with 6 or 12 notches for hexagonal nuts and bolt heads. The 8- and 12-notched sockets are recommended because they will last longer, turn in smaller angular increments, and are about the same price. Socket sizes usually range from 1/8″ to 9/16″ for 1/4″ drives; from 3/8″ to 3/4″ for 3/8″ drives; from 3/8″ to 1-1/4″ for 1/2″ drives; from 3/4″ to 2-1/4″ for 3/4″ drives; and from 1-1/16″ to 3-1/8″ for 1″ drives. Metric sizes are also available.

You may purchase both regular sockets and deep sockets that are especially handy for spinning nuts onto long bolts because they allow part of the bolt into the socket. Sockets with conventional, Phillips head, hex, open-end and box drivers are also available. Socket adapters (Figure 8-8f) are attached to a socket wrench drive when it is desirable to change from one square drive size to another. Socket adapters are available to change socket wrench drives of (in inches): 3/8 to 1/4; 1/4 to 3/8; 3/8 to 1/2; 1/2 to 3/8; 1/2 to 3/4; 3/4 to 1/2; 1 to 3/4 and 3/4 to 1.

A universal joint (Figure 8-8h) makes it possible to turn nuts or bolts where a straight wrench could not be used because of lack of clearance. One end of the joint is attached to a socket wrench drive; the other end attaches to the square hole end of a socket. The joint swings up to 90°. Spring tension holds the joint at any desired angle. Universal joints are also available having 1/2″ square drive with 12-point sockets of 1/2″, 9/16″, 5/8″, 11/16″ and 3/4″. These fixed-size universal joint sockets are useful for applications where the same-sized universal joint is required for extended work.

Extensions (Figure 8-8g), which are connected between the drive and the socket, increase the length of socket wrench drives.

To use a socket wrench, first select the shape (for square or hex nut or bolt) and the size of the socket that is required. Try the socket for proper fit. Next, inspect the clearance area to determine which type of drive is the most satisfactory for the job. If the ratchet drive is selected, snap the socket onto the square drive. Flip the reversing lever to the applicable position for ratcheting in either a clockwise or counterclockwise direction. Use the ratchet wrench speeder to quickly spin the nut or bolt finger-tight. Finally, apply force to the handle until the nut is taut (Figure 8-9). It may be necessary to swing the handle back and forth, but the socket does not have to be

Fig. 8-9. This socket wrench consists of a deep socket and reversible ratchet drive.

removed from the nut. In loosening, reverse the process.

If the flex handle is selected, snap the proper socket onto the square drive. With the drive in a straight line with the handle, spin the handle with the fingers until the nut or bolt is snug. Turn the handle 90° (parallel to the workpiece) and apply pressure to the handle to tighten the fastener.

The T-handle or L-handle drive is used because of its sliding variable handle length. The speeder is used in the same manner as a brace. One hand holds the rear handle and applies a force toward the drive; the other hand rotates the speeder, causing the nut or bolt to be tightened or loosened.

8-15. Strap Wrench

The strap wrench is used for rotating a round object such as a chrome pipe without doing any damage to the surface. The wrench handle (approximately 12″ long) is made of cast iron. The strap, generally made of canvas, has a capacity of 1/8″ to 2″ and can be replaced when worn.

Place the workpiece within the strap loop (Figure 8-1m). Pull the strap tight and apply force to the handle. In Figure 8-1m, force is correctly applied to the handle in a downward direction.

8-16. Torque Wrench

A torque wrench is used to tighten nuts, bolts, capscrews, etc., to specified pressures. The torque wrench is used on machines and automobile and airplane engines when bolts or nuts are to be tightened to the same specified force to assure alignment and equal pressure for sealing. For example, the automobile mechanic would use a torque wrench when installing cylinder heads and main and connecting rod bearing caps.

The two types of torque wrenches are torque-limiting and the deflecting beam-converging scale. Each may incorporate one or more of the following features: square drives of 1/4″, 3/8″, 1/2″, 3/4″ or 1″; quick-release push button for releasing the various English and metric sockets (Section 8-14); ratchet and non-ratchet drives; reversing levers to change the ratchet direction for tightening or loosening; torque measurement in one or both directions; and calibration in a National Bureau of Standards Scale. These torque wrenches are available in calibrated units of inch-pounds and foot-pounds. Handle lengths vary from 13″ to 40″.

The torque-limiting wrench is set to a predetermined specified torque by rotating the micrometer-like handle to the required value as indicated on the scale. A snap lock holds the setting. The nut or bolt to be tightened is then turned. When the predetermined torque pressure is reached, you feel and hear a click in the handle. Since it is not necessary to read the scale the wrench is ideal for use in the dark, in blind spots and even under water or oil. The wrench automatically resets for the next operation.

On deflecting beam-converging scale torque wrenches, the torque pressure value is read out on the scale as the nut or bolt is torqued. If the simple sound device is set for a specified torque value, there is also an audible sound when it is reached.

Torque wrenches are operated in the same way as the socket wrench which is composed of a ratchet drive and socket. First, set the specified torque value into the wrench; next, select the proper-sized socket and snap it onto the torque wrench drive. Place the ratchet reversing lever, if there is one, in the proper position for tightening. Then place the socket fully over the head of the nut or bolt and tighten the fastener until the proper torque pressure is indicated either by an audible click or on the gauge.

Appendix

Appendix A English to Metric and Metric to English Conversion Factors

To Convert	*Into*	*Multiply by*
	English to Metric	
inches	millimeters	25.40
inches	centimeters	2.540
inches	meters	0.0254
feet	millimeters	304.8
feet	centimeters	30.48
feet	meters	0.3048
cubic feet	cubic meters	0.02832
cubic inches	cubic meters	1.639×10^{-5}
square inches	square millimeters	645.2
	Metric to English	
millimeters	inches	0.03937
millimeters	feet	3.281×10^{-3}
centimeters	inches	0.3937
centimeters	feet	3.281×10^{-2}
meters	inches	39.37
meters	feet	3.281
cubic meters	cubic feet	35.31
cubic meters	cubic inches	61,023.0
square meters	square feet	10.76
square millimeters	square inches	1.550×10^{-3}

Appendix B Inch-Millimeter Equivalents of Decimal and Common Fractions

Inch	½'s	¼'s	8ths	16ths	32nds	64ths	Millimeters	Decimals of an Inch[a]
						1	0.397	0.015 625
					1	2	0.794	0.031 25
						3	1.191	0.046 875
				1	2	4	1.588	0.062 5
						5	1.984	0.078 125
					3	6	2.381	0.093 75
						7	2.778	0.109 375
			1	2	4	8	3.175[a]	0.125 0
						9	3.572	0.140 625
					5	10	3.969	0.156 25
						11	4.366	0.171 875
				3	6	12	4.762	0.187 5
						13	5.159	0.203 125
					7	14	5.556	0.218 75
						15	5.953	0.234 375
		1	2	4	8	16	6.350[a]	0.250 0
						17	6.747	0.265 625
					9	18	7.144	0.281 25
						19	7.541	0.296 875
				5	10	20	7.938	0.312 5
						21	8.334	0.328 125
					11	22	8.731	0.343 75
						23	9.128	0.359 375
			3	6	12	24	9.525[a]	0.375 0
						25	9.922	0.390 625
					13	26	10.319	0.406 25
						27	10.716	0.421 875
				7	14	28	11.112	0.437 5
						29	11.509	0.453 125
					15	30	11.906	0.468 75
						31	12.303	0.484 375
	1	2	4	8	16	32	12.700[a]	0.500 0
						33	13.097	0.515 625
					17	34	13.494	0.531 25
						35	13.891	0.546 875
				9	18	36	14.288	0.562 5

Inch	*½'s*	*¼'s*	*8ths*	*16ths*	*32nds*	*64ths*	*Millimeters*	*Decimals of an Inch*[a]
						37	14.684	0.578 125
					19	38	15.081	0.593 75
						39	15.478	0.609 375
			5	10	20	40	15.875[a]	0.625 0
						41	16.272	0.640 625
					21	42	16.669	0.656 25
						43	17.066	0.671 875
				11	22	44	17.462	0.687 5
						45	17.859	0.703 125
					23	46	18.256	0.718 75
						47	18.653	0.734 375
		3	6	12	24	48	19.050[a]	0.750 0
						49	19.447	0.765 625
					25	50	19.844	0.781 25
						51	20.241	0.796 875
				13	26	52	20.638	0.812 5
						53	21.034	0.828 125
					27	54	21.431	0.843 75
						55	21.828	0.859 375
			7	14	28	56	22.225[a]	0.875 0
						57	22.622	0.890 625
					29	58	23.019	0.906 25
						59	23.416	0.921 875
				15	30	60	23.812	0.937 5
						61	24.209	0.953 125
					31	62	24.606	0.968 75
						63	25.003	0.984 375
1	2	4	8	16	32	64	25.400[a]	1.000 0

[a] Exact.

Appendix C Decimal Equivalents of Millimeters

mm.	*Inches*	*mm.*	*Inches*	*mm.*	*Inches*	*mm.*	*Inches*	*mm.*	*Inches*
0.01	0.00039	0.41	0.01614	0.81	0.03189	21	0.82677	61	2.40157
0.02	0.00079	0.42	0.01654	0.82	0.03228	22	0.86614	62	2.44094
0.03	0.00118	0.43	0.01693	0.83	0.03268	23	0.90551	63	2.48031
0.04	0.00157	0.44	0.01732	0.84	0.03307	24	0.94488	64	2.51968
0.05	0.00197	0.45	0.01772	0.85	0.03346	25	0 98425	65	2.55905
0.06	0.00236	0.46	0.01811	0.86	0.03386	26	1.02362	66	2.59842
0.07	0.00276	0.47	0.01850	0.87	0.03425	27	1.06299	67	2.63779
0.08	0.00315	0.48	0.01890	0.88	0.03465	28	1.10236	68	2.67716
0.09	0.00354	0.49	0.01929	0.89	0.03504	29	1.14173	69	2.71653
0.10	0.00394	0.50	0.01969	0.90	0.03543	30	1.18110	70	2.75590
0.11	0.00433	0.51	0.02008	0.91	0.03583	31	1.22047	71	2.79527
0.12	0.00472	0.52	0.02047	0.92	0.03622	32	1.25984	72	2.83464
0.13	0.00512	0.53	0.02087	0.93	0.03661	33	1.29921	73	2.87401
0.14	0.00551	0.54	0.02126	0.94	0.03701	34	1.33858	74	2.91338
0.15	0.00591	0.55	0.02165	0.95	0.03740	35	1.37795	75	2.95275
0.16	0.00630	0.56	0.02205	0.96	0.03780	36	1.41732	76	2.99212
0.17	0.00669	0.57	0.02244	0.97	0.03819	37	1.45669	77	3.03149
0.18	0.00709	0.58	0.02283	0.98	0.03858	38	1.49606	78	3.07086
0.19	0.00748	0.59	0.02323	0.99	0.03898	39	1.53543	79	3.11023
0.20	0.00787	0.60	0.02362	1.00	0.03937	40	1.57480	80	3.14960
0.21	0.00827	0.61	0.02402	1	0.03937	41	1.61417	81	3.18897
0.22	0.00866	0.62	0.02441	2	0.07874	42	1.65354	82	3.22834
0.23	0.00906	0.63	0.02480	3	0.11811	43	1.69291	83	3.26771
0.24	0.00945	0.64	0.02520	4	0.15748	44	1.73228	84	3.30708
0.25	0.00984	0.65	0.02559	5	0.19685	45	1.77165	85	3.34645
0.26	0.01024	0.66	0.02598	6	0.23622	46	1.81102	86	3.38582
0.27	0.01063	0.67	0.02638	7	0.27559	47	1.85039	87	3.42519
0.28	0.01102	0.68	0.02677	8	0.31496	48	1.88976	88	3.46456
0.29	0.01142	0.69	0.02717	9	0.35433	49	1.92913	89	3.50393
0.30	0.01181	0.70	0.02756	10	0.39370	50	1.96850	90	3.54330
0.31	0.01220	0.71	0.02795	11	0.43307	51	2.00787	91	3.58267
0.32	0.01260	0.72	0.02835	12	0.47244	52	2.04724	92	3.62204
0.33	0.01299	0.73	0.02874	13	0.51181	53	2.08661	93	3.66141
0.34	0.01339	0.74	0.02913	14	0.55118	54	2.12598	94	3.70078
0.35	0.01378	0.75	0.02953	15	0.59055	55	2.16535	95	3.74015
0.36	0.01417	0.76	0.02992	16	0.62992	56	2.20472	96	3.77952
0.37	0.01457	0.77	0.03032	17	0.66929	57	2.24409	97	3.81889
0.38	0.01496	0.78	0.03071	18	0.70866	58	2.28346	98	3.85826
0.39	0.01535	0.79	0.03110	19	0.74803	59	2.32283	99	3.89763
0.40	0.01575	0.80	0.03150	20	0.78740	60	2.36220	100	3.93700

Appendix D English System of Weights and Measures

Linear Measure (Length)

1000 mils = 1 inch (in.)
12 inches = 1 foot (ft.)
3 feet = 1 yard (yd.)
5280 feet = 1 mile

Square Measure (Area)

144 square inches (sq.in.) = 1 square foot (sq.ft.)
9 square feet = 1 square yard (sq.yd.)

Cubic Measure (Volume)

1728 cubic inches (cu.in.) = 1 cubic foot (cu.ft.)
27 cubic feet = 1 cubic yard (cu.yd.)
231 cubic inches = 1 U.S. gallon (gal.)
277.27 cubic inches = 1 British imperial gallon (i.gal.)

Liquid Measure (Capacity)

4 fluid ounces (fl.oz.) = 1 gill (gi.)
2 pints = 1 quart (qt.)
4 quarts = 1 gallon

Dry Measure (Capacity)

2 pints = 1 quart
8 quarts = 1 peck (pk.)

Weight (Avoirdupois)

27.3438 grains = 1 dram (dr.)
16 drams = 1 ounce (oz.)
16 ounces = 1 pound (lb.)
100 pounds = 1 hundredweight (cwt.)
112 pounds = 1 long hundredweight (l.cwt.)
2000 pounds = 1 short ton (S.T.)
2240 pounds = 1 long ton (L.T.)

Weight (Troy)

24 grains = 1 pennyweight (dwt.)
20 pennyweights = 1 ounce (oz.t.)
12 ounces = 1 pound (lb.t.)

Angular or Circular Measure

60 seconds = 1 minute
60 minutes = 1 degree
57.2958 degrees = 1 radian
90 degrees = 1 quadrant or right angle
360 degrees = 1 circle or circumference

Appendix E Metric System of Weights and Measures

Linear Measure (Length)

1/10 meter = 1 decimeter (dm)
1/10 decimeter = 1 centimeter (cm)
1/10 centimeter = 1 millimeter (mm)
1/1000 millimeter = 1 micron (μ)
1/1000 micron = 1 millimicron (mμ)
10 meters = 1 dekameter (dkm)
10 dekameters = 1 hectometer (hm)
10 hectometers = 1 kilometer (km)
10 kilometers = 1 myriameter

Square Measure (Area)

1 are = 1 square dekameter (dkm^2)
1 centare = 1 square meter (m^2)
1 hectare = 1 square hectometer (hm^2)

Cubic Measure (Volume)

1 stere = 1 cubic meter (m^3)
1 decistere = 1 cubic decimeter (dm^3)
1 centistere = 1 cubic centimeter (cm^3)
1 dekastere = 1 cubic dekameter (dkm^3)

Capacity

1/10 liter = 1 deciliter (dl)
1/10 deciliter = 1 centiliter (cl)
1/10 centiliter = 1 milliliter (ml)
10 liters = 1 dekaliter (dkl)
100 liters = 1 hectoliter (hl)
1000 liters = 1 kiloliter (kl)
1 kiloliter = 1 stere (s)

Weight

1/10 gram = 1 decigram (dg)
1/10 decigram = 1 centigram (cg)
1/10 centigram = 1 milligram (mg)
10 grams = 1 dekagram (dkg)
100 grams = 1 hectogram (hg)
1000 grams = 1 kilogram (kg)
10,000 grams = 1 myriagram
100,000 grams = 1 quintal (q)
1,000,000 grams = 1 metric ton (t)

Appendix F Conversion Between English and Metric Units

English to Metric	*Metric to English*

Units of Length

1 millimeter = 0.03937 inch or about 1/25 inch

English to Metric	Metric to English
1 inch = 2.540 centimeters	1 centimeter = 0.3937 inch
1 foot = 0.3048 meter	1 decimeter = 3.937 inches
1 yard = 0.9144 meter	1 meter = 39.37 inches
1 mile = 1.6093 kilometers	= 3.281 feet
	= 1.094 yards
	1 kilometer = 0.62137 mile

Units of Area

English to Metric	Metric to English
1 sq. inch = 6.4516 sq. centimeters	1 sq. centimeter = 0.1549997 sq. inch
1 sq. foot = 0.0929 sq. meter	1 sq. meter = 10.764 sq. feet
1 sq. mile = 2.590 sq. kilometers	1 sq. kilometer = 0.3861 sq. mile
1 acre = 0.4047 hectare	1 hectare = 2.471 acres

Units of Volume

English to Metric	Metric to English
1 cu. inch = 16.387 cu. centimeters	1 cu. centimeter = 0.061023 cu. inch
1 cu. foot = 0.028317 cu. meter	1 cu. meter = 35.31445 cu. feet

Capacity (Liquid)

English to Metric	Metric to English
1 gill = 0.11829 liter	1 liter = 8.4537 gills
1 pint = 0.4732 liter	1 liter = 2.1134 pints
1 quart = 0.9463 liter	1 liter = 1.0567 quarts

Capacity (Dry)

English to Metric	Metric to English
1 pint = 0.5506 liter	1 liter = 1.816 pints
1 quart = 1.1012 liters	1 liter = 0.908 quart
1 peck = 8.8096 liters	1 liter = 0.1135 peck
1 bushel = 3.52383 dekaliters	1 dekaliter = 0.28378 bushel

Units of Mass

English to Metric	Metric to English
1 grain = 0.0648 gram	1 gram = 15.432 grains
1 ounce (avdp.) = 28.3495 grams	1 kilogram = 35.274 oz. avdp.
1 pound (avdp.) = 0.45359 kilogram	1 kilogram = 2.2046 lbs. avdp.
1 short ton (2000 lb.) = 0.9072 metric ton	1 metric ton = 1.1023 short tons
1 long ton (2240 lb.) = 1.016 metric tons	1 metric ton = 0.9842 long ton

Appendix G Metric Conversion Table

Millimeters	× 0.03937	= Inches
Millimeters	= 25.400	× Inches
Meters	× 3.2809	= Feet
Meters	= 0.3048	× Feet
Kilometers	× 0.621377	= Miles
Kilometers	= 1.6093	× Miles
Square centimeters	× 0.15500	= Square inches
Square centimeters	= 6.4515	× Square inches
Square meters	× 10.76410	= Square feet
Square meters	= 0.09290	× Square feet
Cubic centimeters	× 0.061025	= Cubic inches
Cubic centimeters	= 16.3866	× Cubic inches
Cubic meters	× 35.3156	= Cubic feet
Cubic meters	= 0.02832	× Cubic feet
Cubic meters	× 1.308	= Cubic yards
Cubic meters	= 0.765	× Cubic yards
Liters	× 61.023	= Cubic inches
Liters	= 0.01639	× Cubic inches
Liters	× 0.26418	= U.S. gallons
Liters	= 3.7854	× U.S. gallons
Grams	× 15.4324	= Grains
Grams	= 0.0648	× Grains
Grams	× 0.03527	= Ounces, avoirdupois
Grams	= 28.3495	× Ounces, avoirdupois
Kilograms	× 2.2046	= Pounds
Kilograms	= 0.4536	× Pounds
Kilograms per square centimeter	× 14.2231	= Pounds per square inch
Kilograms per square centimeter	= 0.0703	× Pounds per square inch
Kilograms per cubic meter	× 0.06243	= Pounds per cubic foot
Kilograms per cubic meter	= 16.01890	× Pounds per cubic foot
Metric tons (1,000 kilograms)	× 1.1023	= Tons (2,000 pounds)
Metric tons	= 0.9072	× Tons (2,000 pounds)
Calories	× 3.9683	= B.T. units
Calories	= 0.2520	× B.T. units

Appendix H Lumber Conversion Chart

Lineal Feet to Board Feet
Example: 1 x 2 x 10 = 1-2/3 bd. ft.

	Length in (Feet)							
Size	*10*	*12*	*14*	*16*	*18*	*20*	*22*	*24*
1 x 2	1-2/3	2	2-1/3	2-2/3	3	3-1/3	3-2/3	4
1 x 3	2½	3	3½	4	4½	5	5½	6
1 x 4	3-1/3	4	4-2/3	5-1/3	6	6-2/3	7-1/3	8
1 x 5	4-1/6	5	5-5/6	6-2/3	7½	8-1/3	9-1/6	10
1 x 6	5	6	7	8	9	10	11	12
1 x 7	5-5/6	7	8-1/6	9-1/3	10½	11-2/3	12-5/6	14
1 x 8	6-2/3	8	9-1/3	10-2/3	12	13-1/3	14-2/3	16
1 x 9	7½	9	10½	12	13½	15	16½	18
1 x 10	8-1/3	10	11-2/3	13-1/3	15	16-2/3	18-1/3	20
1 x 12	10	12	14	16	18	20	22	24
1 x 14	11-2/3	14	16-1/3	18-2/3	21	23-1/3	25-2/3	28
1 x 16	13-1/3	16	18-2/3	21-1/3	24	26-2/3	29-1/3	32
1¼ x 4	4-1/6	5	5-5/6	6-2/3	7½	8-1/3	9-1/6	10
1¼ x 5	5-5/24	6¼	7-7/24	8-1/3	9-3/8	10-5/12	11-11/24	12½
1¼ x 6	6¼	7½	8¾	10	11¼	12½	13-3/4	15
1¼ x 8	8-1/3	10	11-2/3	13-1/3	15	16-2/3	18-1/3	20
1¼ x 9	9-3/8	11¼	13-1/8	15	16-7/8	18¾	20-5/8	22½
1¼ x 10	10-5/12	12½	14-7/12	16-2/3	18¾	20-5/6	22-11/12	25
1¼ x 12	12½	15	17½	20	22½	25	27½	30
2 x 2	3-1/3	4	4-2/3	5-1/3	6	6-2/3	7-1/3	8
2 x 3	5	6	7	8	9	10	11	12
2 x 4	6-2/3	8	9-1/3	10-2/3	12	13-1/3	14-2/3	16
2 x 6	10	12	14	16	18	20	22	24
2 x 8	13-1/3	16	18-2/3	21-1/3	24	26-2/3	29-1/3	32
2 x 9	15	18	21	24	27	30	33	36
2 x 10	16-2/3	20	23-1/3	26-2/3	30	33-1/3	36-2/3	40
2 x 12	20	24	28	32	36	40	44	48

Appendix I Nail Reference Chart

COMMON WIRE NAILS

Size	*Length*	*Gauge*	*Approx. No. to lb.*
2D	1 In.	No. 15	876
3D	1¼	14	568
4D	1½	12½	316
5D	1-3/4	12½	271
6D	2	11½	181
7D	2¼	11½	161
8D	2½	10¼	106
9D	2-3/4	10¼	96

Size	*Length*	*Gauge*	*Approx. No. to lb.*
10D	3 In.	No. 9	69
12D	3¼	9	63
16D	3½	8	49
20D	4	6	31
30D	4½	5	24
40D	5	4	18
50D	5½	3	14
60D	6	2	11

FLOORING BRADS

Size	*Length*	*Gauge*	*Approx. No. to lb.*
6D	2 In.	No. 11	157
7D	2¼	11	139
8D	2½	10	99
9D	2-3/4	10	90
10D	3	9	69
12D	3¼	8	54
16D	3½	7	43
20D	4	6	31

FINISHING NAILS

Size	*Length*	*Gauge*	*Approx. No. to lb.*
2D	1 In.	No. 16½	1351
3D	1¼	15½	807
4D	1½	15	584
5D	1-3/4	15	500
6D	2	13	309
7D	2¼	13	238
8D	2½	12½	189
9D	2-3/4	12½	172
10D	3	11½	121
12D	3¼	11½	113
16D	3½	11	90
20D	4	10	62

CASING NAILS

Size	*Length*	*Gauge*	*Approx. No. to lb.*
2D	1 In.	No. 15½	1010
3D	1¼	14½	635
4D	1½	14	473
5D	1-3/4	14	406
6D	2	12½	236
7D	2¼	12½	210
8D	2½	11½	145
9D	2-3/4	11½	132
10D	3	10½	94
12D	3¼	10½	87
16D	3½	10	71
20D	4	9	52
30D	4½	9	46

SMOOTH AND BARBED BOX NAILS

Size	*Length*	*Gauge*	*Approx. No. to lb.*
2D	1 In.	No. 15½	1010
3D	1¼	14½	635
4D	1½	14	473
5D	1-3/4	14	406
6D	2	12½	236
7D	2¼	12½	210
8D	2½	11½	145
9D	2-3/4	11½	132
10D	3	10½	94
12D	3¼	10½	88
16D	3½	10	71
20D	4	9	52
30D	4½	9	46
40D	5	8	35

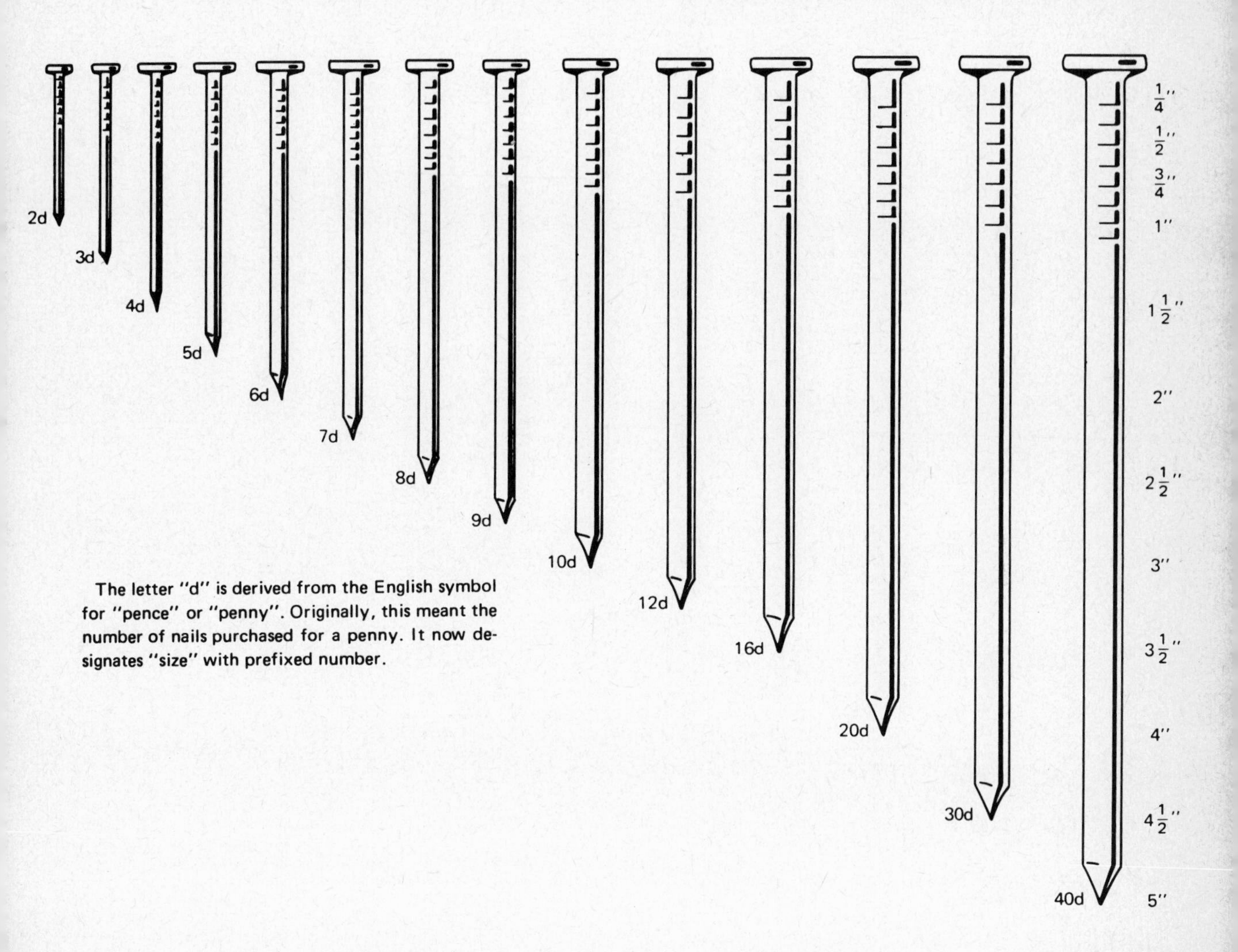

The letter "d" is derived from the English symbol for "pence" or "penny". Originally, this meant the number of nails purchased for a penny. It now designates "size" with prefixed number.

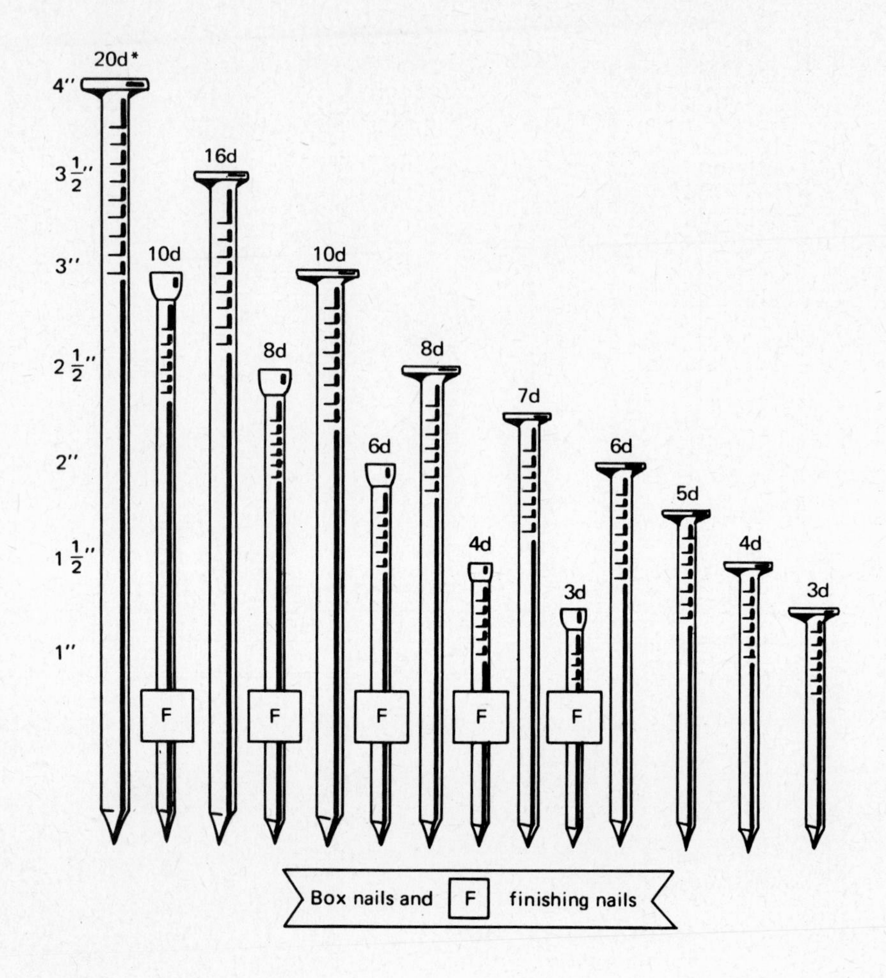

Wire nails and brads			
Length	Gauges	Length	Gauges
3/16"	20 to 24	1"	7 to 20
1/4"	19 to 26	1 1/8"	7 to 19
3/8"	18 to 26	1 1/4"	6 to 17
1/2"	14 to 24	1 3/8"	6 to 17
5/8"	12 to 24	1 1/2"	4 to 17
3/4"	10 to 21	1 5/8"	4 to 17
7/8"	8 to 20	1 3/4"	4 to 17

Appendix J Screw Reference Chart

Listed below are screw lengths from $\frac{1}{4}$" to 4". Shank dimensions are shown from 0 to 24.
These sizes are most frequently used and are more generally available.

Length	Shank numbers																	
1/4 inch	0	1	2	3														
3/8 inch			2	3	4	5	6	7										
1/2 inch			2	3	4	5	6	7	8									
5/8 inch				3	4	5	6	7	8	9	10							
3/4 inch					4	5	6	7	8	9	10	11						
7/8 inch							6	7	8	9	10	11	12					
1 inch							6	7	8	9	10	11	12	14				
1 1/4 inch								7	8	9	10	11	12	14	16			
1 1/2 inch							6	7	8	9	10	11	12	14	16	18		
1 3/4 inch									8	9	10	11	12	14	16	18	20	
2 inch									8	9	10	11	12	14	16	18	20	
2 1/4 inch										9	10	11	12	14	16	18	20	
2 1/2 inch													12	14	16	18	20	
2 3/4 inch														14	16	18	20	
3 inch															16	18	20	
3 1/2 inch																18	20	24
4 inch																18	20	24
0 to 24 diameter dimensions in inches at body	0.060	0.073	0.086	0.099	0.112	0.125	0.138	0.151	0.164	0.177	0.190	0.203	0.216	0.242	0.268	0.294	0.320	0.372
Twist bit sizes for round, flat and oval head screws in drilling shank and pilot holes.																		
Shank holes hard and softwood	1/16	5/64	3/32	7/64	7/64	1/8	9/64	5/32	11/64	3/16	3/16	13/64	7/32	1/4	17/64	19/64	21/64	3/8
Pilot hole softwood	1/64	1/32	1/32	3/64	3/64	1/16	1/16	1/16	5/64	5/64	3/32	3/32	7/64	7/64	9/64	9/64	11/64	3/16
Pilot hole hardwood	1/32	1/32	3/64	1/16	1/16	5/64	5/64	3/32	3/32	7/64	7/64	1/8	1/8	9/64	5/32	3/16	3/64	7/32
Auger bit sizes for countersunk heads			3	4	4	4	5	5	6	6	6	7	7	8	9	10	11	12

Twist drill sizes for round, oval and flat head screws in drilling shank and pilot holes.																			
Screw sizes—common, slotted head		0	1	2	3	4	5	6	7	8	9	10	11	12	14	16	18	20	24
Shank hole—hard and softwood	Fractional	1/16	5/64	3/32	7/64	7/64	1/8	9/64	5/32	11/64	3/16	3/16	13/64	7/32	1/4	17/64	19/64	21/64	3/8
	number- letter	52	47	42	37	32	30	27	22	18	14	10	4	2	D	I	N	P	V
Pilot hole—softwood	Fractional	1/64	1/32	1/32	3/64	3/64	1/16	1/16	1/16	5/64	5/64	3/32	3/32	7/64	7/64	9/64	9/64	11/64	3/16
	number	75	71	65	58	55	53	52	51	48	45	43	40	38	32	29	26	19	15
Pilot hole—hardwood	Fractional	1/32	1/32	3/64	1/16	1/16	5/64	5/64	3/32	3/32	7/64	7/64	1/8	1/8	9/64	5/32	3/16	13/64	7/32
	number	70	66	56	54	52	49	47	44	40	37	33	31	30	25	18	13	4	1

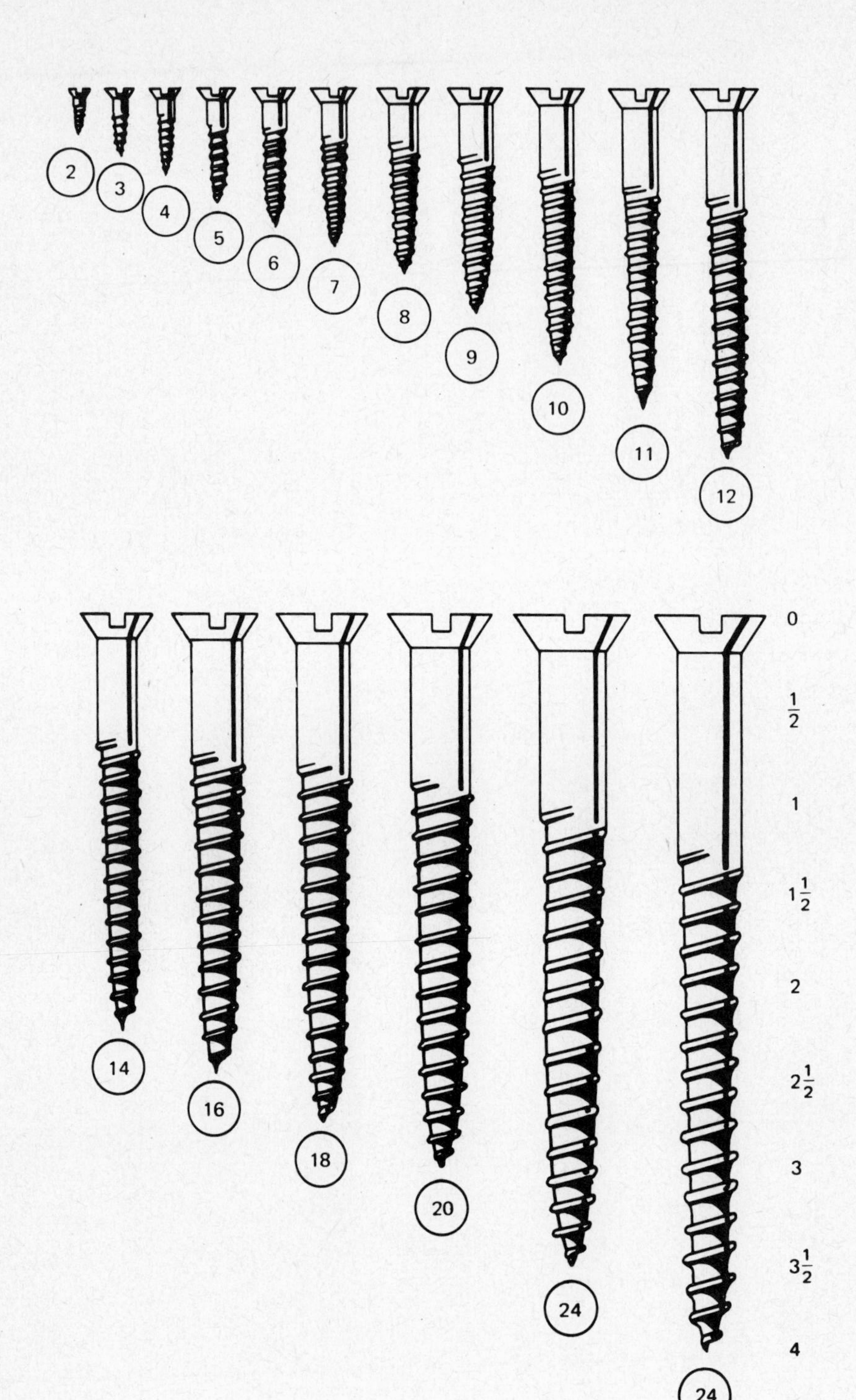
2
3
4
5
6
7
8
9
10
11
12
14
16
18
20
24
24
0
1/2
1
1 1/2
2
2 1/2
3
3 1/2
4

Appendix K Decimal Equivalents of Number and Letter Size Drills

NUMBER SIZE DRILLS

No.	*Size of Drill In Inches*	*No.*	*Size of Drill In Inches*	*No.*	*Size of Drill In Inches*	*No.*	*Size of Drill In Inches*
1	.2280	21	.1590	41	.0960	61	.0390
2	.2210	22	.1570	42	.0935	62	.0380
3	.2130	23	.1540	43	.0890	63	.0370
4	.2090	24	.1520	44	.0860	64	.0360
5	.2055	25	.1495	45	.0820	65	.0350
6	.2040	26	.1470	46	.0810	66	.0330
7	.2010	27	.1440	47	.0785	67	.0320
8	.1990	28	.1405	48	.0760	68	.0310
9	.1960	29	.1360	49	.0730	69	.0292
10	.1935	30	.1285	50	.0700	70	.0280
11	.1910	31	.1200	51	.0670	71	.0260
12	.1890	32	.1160	52	.0635	72	.0250
13	.1850	33	.1130	53	.0595	73	.0240
14	.1820	34	.1110	54	.0550	74	.0225
15	.1800	35	.1100	55	.0520	75	.0210
16	.1770	36	.1065	56	.0465	76	.0200
17	.1730	37	.1040	57	.0430	77	.0180
18	.1695	38	.1015	58	.0420	78	.0160
19	.1660	39	.0995	59	.0410	79	.0145
20	.1610	40	.0980	60	.0400	80	.0135

LETTER SIZE DRILLS

A	0.234	J	0.277	S	0.348
B	0.238	K	0.281	T	0.358
C	0.242	L	0.290	U	0.368
D	0.246	M	0.295	V	0.377
E	0.250	N	0.302	W	0.386
F	0.257	O	0.316	X	0.397
G	0.261	P	0.323	Y	0.404
H	0.266	Q	0.332	Z	0.413
I	0.272	R	0.339		

Glossary

This glossary lists terms which are used in this series and which may be unfamiliar to you. Definitions and descriptions of specific tools are contained throughout the book.

Acute – being or forming an angle less than 90°.

Anvil – the flat end of a tool, such as a punch or chisel, that is struck with a hammer to cause the tool to function; the flat part of a machinist's vise; a heavy steel-faced iron block on which metal is shaped.

Arbor – a shaft on which a revolving cutting tool is mounted.

Bead – a projecting rim, band or molding.

Bevel Cut – a cut made on one surface at an angle (except a right angle) to another surface; oblique; also a tool for making angles.

Burr – a rough protuberance, ridge or area left on metal after cutting or drilling; to form a rough point or edge on.

Casein – a phosphoprotein of milk: as one that is precipitated from milk by heating with an acid or by the action of lactic acid in souring; used in making paints and adhesives.

Catalysis – a modification in the rate of a chemical reaction induced by material unchanged chemically at the end of the reaction.

Catalyst – a substance that initiates a chemical reaction and enables it to proceed under milder conditions than otherwise possible.

Catalytic – causing, involving or relating to catalysis.

Chamfer – an oblique surface, usually 45°, cut on the edge or corner of a board.

Chuck – an attachment for holding a workpiece or tool in a machine (as in an electric drill).

Collet – a metal band, collar, ferrule or flange; used to hold small diameter bits.

Concave – hollowed or rounded inward like the inside of a bowl.

Convex – curved like a circle or sphere when viewed from without; bulging and curved.

Counterbore – to enlarge the upper part of a hole to receive and allow the head of a screw or bolt to be recessed below the surface.

Countersink – to enlarge the upper part of a hole by chamfering to receive the cone-shaped head of a screw or bolt.

Dado – to cut a rectangular groove into a workpiece; the blade used to cut the rectangular groove; a rectangular groove cut into a workpiece.

Dog – the retractable bar on a wood vise used with a bench stop or other device to clamp a workpiece.

Dovetail – a joint or fastening formed by one or more tenons and mortises spread in the shape of a dove's tail.

End Grain – the end grain of a board is the grain at the end of a board in a direction that is across the grain of the wood. See grain.

Ferrous Metal – a metal containing iron.

Flats – (drill shank flats, chuck flats) – the flat areas on the end of some drill shanks (power wood-boring bits) that are placed against the flats in the drill chuck. These flats align the bit into the chuck and also prevent the shank form slipping in the chuck.

Flux – a substance used to promote fusion, especially of metals.

Grain – (of wood) – the stratification of the wood fibers in a piece of wood. Cutting *with* *the grain* is the process of cutting in the direction of the stratification.

Hasp – a clasp for a door or lid, especially one passing over a staple and fastened by a pin or padlock.

Helical – having the form of a helix.

Helix – something spiral in form; a curve traced on a cylinder by the rotation of a point crossing its right sections at a constant oblique angle.

Hone – a fine grit stone for sharpening a cutting implement; to sharpen, enlarge, or smooth with a hone.

Kerf – the cut or incision made by a saw or other instrument.

Keyway – a groove or channel for a key or spline.

Kiln – a furnace or oven used to dry something such as wood, brick or ceramic.

Laminate – to make by uniting superimposed layers of one or more materials such as counter topping on wood counters, fabric on wood, etc.

Lateral – of or relating to the side; situated on, directed toward, or coming from the side.

Layout – the marking off of points, lines, circles, arcs and angles on metal or wood as a guide for cutting, matching parts, drilling, etc.

Malleable – capable of being extended or shaped by beating with a hammer or by the pressure of rollers.

Meter – the meter is the fundamental unit of length of the metric system and is equal to 39.37 inches or 3.281 feet. The meter is measured in cadmium – red light waves and is equal to 1,553,164.13 of the waves.

Miter – the abutting surface or bevel on either of the pieces joined in a miter joint. A *mitered joint* is a joint formed when two pieces of identical cross section are joined at the ends, and where the joined ends are beveled at equal angles.

Mortise – a rectangular cut of considerable depth in a piece of wood for receiving a corresponding projection (tenon) on another piece of wood to form a joint.

Mushrooming – the spreading out of the metal of a tool when it is hammered; as the head of a chisel spreading out from repeated hammering during chisel cutting operations.

Ogee – a molding with an S-shaped profile.

Pantograph – an instrument for copying on a predetermined scale; it consists of four light, rigid bars joined in parallelogram form.

Parallax – the apparent displacement of the reading mark or graduation on a scale due to a change of direction in the position of the observer.

Parallel – lying in the same plane but never meeting no matter how far extended.

Pawl – a pivoted tongue or sliding bolt on one part of a machine that is adapted to fall into notches on another part (as a ratchet wheel) so as to permit motion in only one direction.

Perpendicular – vertical; upright; meeting a given line or surface at right angles.

Pilot Hole – a hole of small diameter drilled into a material to lead the point of a larger drill.

Porous (material) – a material that is permeable to liquids; example: wood fibers allow glue to soak into them.

Quench – to cool rapidly with water.

Rabbet – a cut, groove, or recess made on the edge or surface of a board to receive the end or edge of another board or the like which is similarly shaped.

Ratchet – a mechanism that consists of a wheel having inclined teeth into which a pawl drops so that motion can be imparted to the wheel, governed, or prevented and that is used in a hand tool to allow effective motion in one direction only.

Right Angle – the angle formed by two perpendicular lines intercepting a quarter of a circle drawn about its vertex. A right angle is an angle of 90°.

Scrollwork – ornamental work cut out with a coping saw or power jig or scroll saw.

Shaper – a power tool having cutters used to shape wood workpiece edges, molding.

Shim – a thin strip of metal or wood for filling in, as for bringing one workpiece in line with another.

Solder – a metal or metallic alloy used when melted to join metallic surfaces.

Spline – a thin wood or metal strip used in building construction; a key that is fixed to one of two connected mechanical parts and fits into a keyway in the other.

Strop – a strip of leather or other flexible material used for sharpening razors or tool edges.

Stylus – an instrument for writing, marking or incising.

Tang – a long and slender projecting strip, tongue, or prong forming part of an object as a file or bit, etc. and serving as a means of attachment for another part, as a handle or brace.

Tangent – touching, as a straight line in relation to a curve or surface.

Temper – to soften by reheating at a lower temperature; to harden (steel) by reheating and cooling in oil.

Template – a gauge, pattern, or mold (as a thin plate or board) used as a guide to the form of a piece being made.

Tenon – a projection fashioned on an end of a piece of wood for insertion into a corresponding cavity (mortise) in another piece of wood to form a joint.

Tensile Strength – the greatest longitudinal stress a substance can bear without tearing apart.

Vial – a small glass or acrylic tube with a liquid. The vial in a level is not completely filled; therefore, an air bubble remains that indicates a level condition when centered between marks on the vial.

Viscosity – a property of a fluid which resists change in the shape or arrangement of its elements during flow.

Viscous – sticky, adhesive; of a thick nature; having the property of viscosity.

Vitrify – to convert into glass or a glassy substance by heat and fusion; to become vitrified.

Whetstone – a tool for whetting edge tools.